The R.A.M.S. Library of Alchemy

Volume 10

The Five Centuries

By

Rudolph Glauber

R.A.M.S. Publishing Company

The Five Centuries

By

Rudolph Glauber

Produced by

Restorers of Alchemical Manuscripts Society

R.A.M.S. Publishing Company

R.A.M.S. Publishing Company
117 Rutherford Lane
Stuarts Draft VA 24477

First Edition 2015

ISBN-13 **978-1508686910**

ISBN-10 **1508686912**

Image Processing by Philip N. Wheeler

Printed in the United States of America

Disclaimer

Liability: The publisher does not warrant or assume any legal liability or responsibility for the accuracy, completeness, or usefulness of any information, apparatus, product, or process disclosed. The publisher makes no representation as to the accuracy or completeness of the contents of this book and specifically disclaims any implied warranty of merchantability or fitness for a particular purpose. No warranty may be created or extended by written sales materials or sales representatives. You should obtain professional consultation where appropriate. The publisher shall not be liable for any loss of profit or other commercial or personal damages, including but not limited to special, incidental, consequential, or other damages.

THE COMPLETE WORKS
OF

RUDOLPH
GLAUBER

trans: Chris. Packe

RAMS
1983

THE CENTURYS

FIRST

Table of Contents

Dedicated to Hans W. Nintzel,
American Alchemist
and
Founder of the
Restorers of Alchemical Manuscripts Society
(R.A.M.S.)

Introduction

Philip N. Wheeler

Johann Rudolph Glauber (1604 – 1670) was a prominent Dutch-German alchemist and chemist, who has also been called one of the first chemical engineers. Christopher Packe translated the bulk of Glauber's works on Alchemy from High-Dutch into English. His translation was published in London in 1689. Hans Nintzel recognized the importance of this work and included it in the R.A.M.S. Library.

"The Five Centuries" is but a small part of Glauber's works. The rest of Glauber's Alchemical works are also available in The R.A.M.S. Library of Alchemy.

This text was typed by Hans Nintzel and D. H., a frequent contributor to the R.A.M.S. effort. I scanned the typed pages and, using optical character recognition (OCR) software, converted them into digital text. In spite of extensive editing, a few OCR errors might still be found in the text.

THE

SECOND PART

OF

GLAUBER'S WORKS.

THE FIRST CENTURY,

OR

Wealthy Store-house of Treasures
Being

A General APPENDIX to all his hitherto-published
Writings

The which doth not only illustrate all obscure places, as well in his Philosophical and. Medicinal as Chymical Writings, and explain those hard places to be understood; but also do so abundantly supply those which are defective, that the learned and the unlearned, the highest and the lowest, and more, the meanest workman and 1{usbandman, may sufficiently be able to comprehend that GLAUBER hath in all his writings, written the pure and simple Truth, and hath again brought to light the most noble Art of Alchemy which hath so long lain hid in darkness hitherto; and hath discovered it for the common good of Mankind.

LONDON, Printed in the Year MDCLXXXIX

PREFACE

Courteous Reader,

    That every promise becomes a debt, is reported by a common Proverb or Byword, and therefore to stand to ones word or promise, is a thing which procures a great Ornament or grace no less to JUNIORS than to SENIORS. Since therefore in my writings I have ingaged my Faith or Credit through the promising of some little works, and yet have not been able, by reason of the scantiness of time, hitherto to satisfje the expectation and desire of very many, by publishing of the same; yea since greater Discommodities and Impediments being cast in my way, do hinder me from day to day whereby I cannot write more things, although I have nothing more in my desires than that in standing to my promise, I may acquit my credit, and set forth the said little works; to wit, my Vegetable work, my Work of SATURN, my Book of Dialogues or Discourse, the fourth part of my SPAGYRICAL PHARMACOPOEA or Chymical Dispensatory, and my admirable little Book of the concentring of the Heaven and Earth; truly they being Treatises containing most excellent Arcanums or Secrets, and the most worthy ones whereof Men can be made partakers, and notwithstanding I am of necessity destitute of time for the writing of any Treatise peculiarly, and for that cause I am constrained to insist in a nearer path, and for the sake of promoting the publick good, to send forth the said Treatises in publick by a less labour and trouble. The present Treatise notified with the Title of an universal

Chest or Cabinet full of Riches, or of a general Appendix of all my writings hitherto exposed to the publick view performs this, whereby all things which have been either the more briefly and obscurely spoken in them, are with a more clear or perspicuous illustration explained, or things that have been wholly omitted are supplied, and by the same endeavour the promised Treatises are added, yet not in that order wherein they ought otherwise to be written down; and the which order here to be observed, would administer very much trouble; but wherein all the secrets have in process oftime been made known unto me, and committed to Paper. But it is free for any one to add according to his own Judgement, Medicinal Secrets unto Medicinal ones, Mineral Secrets unto Mineral ones, Chymical Secrets unto Chymical ones, if it shall so please him, and time shall also permit the same, which it in no wise permitteth unto me, every one that acquiesceth and is content with these things may consider, if a certain Cook should set a Dish on the Table filled with the best Meats, as being destitute of time, to put every sort of Meat in a several Dish, whether he could of right be angry with him, or by whisperingly prating, he could dare to say, he was to be blamed as being not skillfull in the affairs of the Kitchin, because collecting so many delicate and such dainty Meats into one Dish, and daring to set them on the Table? I suppose not any one could of right complain of such a deed of any Cook; the Cook desiring to have it taken in good part, such Meats as he had, such he sets before them; he that refuseth to take of them, may use his own liberty, and may let those Meats alone, which he is not compelled to receive, even as the Cook also may be constrained by none in preparing of the same according to his own will or judgement.

Whatsoever Meat doth not please the Pallate of one, yet will not be ingratefull to the Pallate of another, but on the contrary grate-full, seeing one Food is wont to savour or relish this Man, and another the other, neither is he inordinately affected with the disdain of confused Meats, who taketh of those which relish him, and leaveth the rest brothers.

Let every one that blameth these writings do the same, not in hastily taking them in evil part, but in friendly and courteously excusing me that I have not sent them abroad in a more harmonious order.

They are like unto a certain true and great Cabinet or Chest, filled with very many excellent Secrets, being reduced into my knowledge through a successive diligent search of thirty years and so collected into one heap, that they might either be conserved for my own or at some time be made of publick use or service; out of this Chest every one shall be able to exhaust those ARCANUMS and SECRETS, which shall please him or serve his uses.

As to what concerns my self, because I daily behold sometimes this Man, sometimes that Man being snatched away by death, to be carried forth and committed to the Earth, I may easily conclude rationally with my self, that those changes or chances will in a short time happen also unto me; I should commit a very grievous offence or errour, that so many costs, labours and troubles, of so many and so great Secrets being consumed in vain, I should carry them away with me underground, and not bestow them for a common good: I shall here perform the office of a good House-holder, or skillfull House-keeper or Steward, whO after that he hath made abundant of Provision for Winter-cloathing for himself, his wife, Children and whole Family, if he hath as yet plenty of Linnen and Woollen Cloth remaining, he doth not cast them away, but rather casts them together into a Chest, so long to be kept, until he shall obtain an occasion of administring them for the use of his Neighbour. In the name of the Lord therefore, in making a beginning with the opening of my Chest of Treasures, I will empty it out by little and little by degrees, and will offer it for a common use, that out of so many Treasures, every one may convert unto his own use, what things he shall judge to be profitable unto him; to wit, a Physician Medicinal things, and a Chymist Chymical things, even as every one shall discern any thing to be fit for his own use, every one of what rank soever shall find those things wherewith he might be content, so indeed that whatsoever he shall not meet withall in the first, second, or third Century may be found in the rest, for which things sake, if ten Centuries shall not be sufficient, I will adjoin other ten or more, that so I may remove from me all those cares, and carefulnesses wherewith the custody of so great Treasures hath importuned and affected me for so many years. Like unto a travelling woman, who with the greatest desire

14

expecteth the hour of her delivery, and who desireth the beholding of her Fruit, do I desire that time wherein all things shall be printed in Letters. The Almighty God bestow on me so much presence of mind, health, and strength, and prolong my Life so far, that I may finish it to his Honour, and the Succour, Comfort and Profit of all Mankind.

AMEN.

THE FIRST CENTURY

OR

GLAUBER'S Wealthy Store-house of Treasures.

PART II.

In the NAME of the most Holy Trinity I begin to write the First Century of my General APPENDIX, or an Exposition of all my Writings hitherto set forth.

I.   Concerning Fire and Salt, and what Alchemy is.

Alchemy is a Science, and Art of destroying, of purging immature or unripe and impure Metals, by Fire and Salt, and by a singular Artifice, of converting the more pure part into a better form and kind, according to the words of PARACELSUS, who saith, EVERY SOMETHING IS TO BE CONVERTED INTO NOTHING, AND EVERY NOTHING INTO SOMETHING. Also Corruption renders that which is good perfect, the which is to be equally understood of particular and universal Operations.

Nevertheless it is not of necessity that I should here tediously treat of a particular Transmutation of Metals by Fire and Salt, because that hath been already long since performed in the second part of the Miracle of the World, and all in the fifth part of the prosperity of GERMANY.

A Square within a Circle.

In the Sun
and Salt are
all things.

II. A Demonstration whereby it is proved that Fire and Salt are most noble Creatures of God, and that in Fire there lies hid the purest Salt, and in Salt a most efficacious Fire.

That Fire and Salt are most noble Creatures of God I have evidently enough demonstrated in my little work concerning the nature of Salts. But that there is a subtle saltish spirit in any fire, and that between the heat of the Sun, and of our Kitchin fire, as to their saltish spirit, a great difference doth interpose is sufficiently and over sufficiently known. But to comprehend or lay hold of, to concenter such a pure saltish spirit of SOL, to render it corporeal, palpable or perceivable, and visible, remains hidden and unknown to us by reason of our sins; because God reserves so great mysteries for his own alone, of whom he is honoured and feared; for God himself useth no better similitude than the fire, whereunto the Ancients exhibited divine honour, and by the help of the same, perfected all their Sacrifices: So among the CALDEANS, Fire, and God are called by one and the same name of ESCH: and among the HEATHENS, the chief Philosophers, yea HERMES himself thought the Sun to be a God, and worshipped it for a God. These things are found expressed by MUTIUS concerning the nature of Gods, and therefore those things are not necessary which may be here repeated. Yet this is well to be noted, that God hath always appeared to his Saints under the shew of fire, and hath talked with them out of it, it being that which is full of the greatest mysteries, yet observed but by a few, as in a peculiar little work concerning the concentration of the Heaven and the Earth I will more plainly and fully declare. I affirm therefore, that it can scarce be, that the admirable, yea incredible nature of fire should be described without the revelation of the highest or greatest mysteries of God. Therefore it is better that such Secrets are passed by in silence, than that precious pearls should be cast before Swine, who, are wont to receive them with laughter, and proclaim that they are nothing but the mere sophisteries of triflers, even as is evidently manifest from the description of J. H. S. of the Philosophers Stone, wherein Nature, he saith, makes not use of Glasses, Vessels, Fire, Salt, Urine, and the like in the bowels of the Earth, and the universal ELIXIR may very fitly be prepared by him, who also hath not handled any Chymical Labours, or was never busied about Fire and Salt: Let it shame the man of such stinking lies, wherewith he endeavours to cover his own ignorance, I on the contrary affirm, that all those that know not how to handle Fire and Salt, do in very deed know nothing, but do give credit unto those things only which they hear, or read in others writings, and also for that cause are unworthy of the name and title of true Philosophers; for true Philosophy is to be thoroughly or perfectly learned by the help of Fire and Salt alone, the which God willing shall be more evidently demonstrated.

III. It is moreover demonstrated, that inall Salts an admirable Fire doth lurk as being laid up therein, though the indeavour whereof very many admirable things may be perfected as well in Medicine as in Alchemy; and also that it may be altogether performed, that out of Vitriol the Stone of the ancient Wise men, out of Salt peter a spiritual Gold, and an excellent yellow tincture, and out of common Salt the true Pearl of the Philosophers may be prepared.

In all Salts, that a most potent Fire doth lurk as being laid up therein, those have best known who have the labours of the fire thoroughly viewed and certainly known. For through the efficacy and operation hereof, salts are reduced unto a fiery force, or power or unto a moist fire, out of which they before arose, after the laying down of their earthiness, yet one salt draws out one fire far unlike to the fire of another, so that this is volatile, the other is fixed, and remaining constant in the fire; another is partly volatile and partly fixed, even as the operation shall procure this or the other property unto them, yet all such fiery salts may by the benefit of Art be concentred, and made more efficacious than they were made by some one distillation. For example sake. If any one beholdeth Vitriol, and considers of the nature thereof, he shall in very deed certainly find that by the help of, a

19

strong fire, there may be allured or extracted out of it that which was in the beginning, to wit, a fiery spirit, which by the aid of external heat or fire, being reduced into a narrow Central room, or Concentration, draws out that internal fire, uncloathing it self of, or displaying so great virtues, that it reduceth into a Coal all things which it moisteneth or encompasseth, even like as if it had been burnt up by common Kitchin fire, or by Glasses receiving the Sun-beams and burning up all things that are objected against it. Concerning these fiery salts, and the preparation and use of them, I being here to deliver a few things, I will take my beginning from the fire of Vitriol, and the preparation of the same, the various and manifold use whereof shall be afterwards explained in its own place.

IV. Of the Preparation of the Fire of Vitriol.

Retorts made of the best earth do draw out the fire of Vitriol by distillation after this manner following.

Common Vitriol is calcined in earthen pots unto a redness, and reduced into a powder, it is put into an earthen Retort, and placed in a Furnace, and a great vessel adjoined to the neck of the Retort, which is to receive the spirits going forth; the fire is kindled by degrees, and gradually increased until the Retort be brightly red hot, in which degree of fire it is so long to be urged until no white Clouds or little vapoury Mists do any longer appear. This operation is perfected in 24 hours space at the most. But if the Retort shall be very large all the Oil cannot be extracted in the space of 24 hours, but will require a longer time for the operation, which experience it self will determine; after all the Spirits are distilled off and settled to the bottom of the Receiver, the clay luting which joined the Receiver to the neck of the Retort is to be mollified with a wet cloth put round about it, and the Receiver taken off, and the spirits poured out of it into a glass body well coated with Clay, the which (having the Alembick put on) is to be set in sand, that the volatile spirit may slowly and gently be drawn off, and kept for its use afterwards to be taught. Also afterwards the phlegm is to be drawn off, and reserved for its own uses, because it hath its own peculiar virtues. At length also the last spirit is to be received in a peculiar vessel the which, after that it hath ceased, and fiery drops do follow, the fire is (by degrees) to be removed, and when the sand is cold, the gourd is to be taken out, in which (the Alembick or head being taken away) thou shalt find a fiery Oil of a black or somewhat reddish colour, the which is again to be rectified in an open fire in a Retort well coated, that it may be rendred more fiery and clear.

By this Oil admirable things, and those not only profitable for Physicians but also for Chymists, and other Artificers, are perfected as we shall straitway see.

There are indeed other ways or means also by which this oil is attained, but this afore taught is the easiest of all, although it require the more time. But if any one stand in need of a greater of the same, he may procute those greater Cans prepared of the best Earth, they being so joined to each other, that the uppermost being placed on the fire, the rest might be placed without the fire, so that the lowermost may receive the oil going out by descent.

V. A proof whether this Oil of Vitriol be well prepared and strong, and fit enough for that operation of which we here treat.

Let down a quill or some small piece of wood into the Oil, the which, when thou hast left in it for some small time, draw it out; if it shall be burnt unto a Coal the oil is well prepared, but if not, it is a sign that some what of moisture is as yet therein, which is again to be expelled by fire.

VI. Another tryal or experiment.

Dip in the oil a piece of woolen, linnen, or which is better a piece of cotton cloth extended to the breadth of a finger, and pour on the same being taken out and laid down some drops of the spirit or oil of Turpentine, the which if being kindled they shall conceive a flame it is a sign that the oil was well prepared.

VII. Another further Proof.

Pour into some little glass some small quantity of spirit of wine wanting all phlegm, and pour on the same some drops of this oil by little and little; and if the spirit of wine kindle and burn all away the oil is prepared after a due manner.

N. B. I admonish that every one doth warily handle this operation; for in these two fires, to wit, saltish and sulphureous ones, there is great virtue hidden, the which seems probable but to a few, if it should be manifested unto them, neither that have I consulted or decreed that it should be made known to very many. These few particulars do sufficiently teach after what sort such fires are to be used in Medicine, Alchemy and other Arts; but these experiments are sufficient.

VIII. Concerning the use of this Fire of Vitriol in Medicine.

The use of this fire, as also of the volatile spirit of the same, and of its flegm, thou shalt find described in the second part of my Furnaces, and among other Authours; so that the repetition thereof is here superfluous, this is only to be known that this fire being only besmeared or anointed with a feather on all uncurable and Cancerous or eating Ulcers, kills the Poison, and causeth that such Ulcers do very easily admit of cure, if so be the Escharre be but first removed by the applying some ointment or emplaister which cures adustion or burning. For this oil burns up all wild or foreign flesh, and that which (as proud) lifts up it self with an abounding poison, like unto a certain bright burning Iron, and separates all evil and hurtfull flesh from the good and sound flesh.

IX. Of the general use of this Oil in Alchemy.

By this mineral fire, all kind of Transmutatjons
of things are perfected, but particularly it exalteth
some of the more base metals into a higher degree, and
makes them more constant, of which more shall be said in
the following Chapters or Treatises.

In the general, some Vegetables, Animal and
Mineral subjects, may by the operation of this Oil be
reduced into fixt Medicines, and indeed far more
commodiously than by the common fire of Wood or Coals.
And moreover which is a far greater thing in this very
oil a fiery Tincture is hidden, and is manifested by the
benefit of Art, as Fryer BASILIUS and other Philosophers
do affirm.

X. Of the use of this fire in other Arts.

By the virtues of this invisible, and yet
essential fire, all sorts of most profitable matters are
performed, the which notwithstanding is not here safe
for me to describe, but I am constrained to refer it
till another time, it only in this place seems worthy my
labour, briefly to shew that this fire performs all
those things which the fire of Coals is otherwise wont
to effect.

Truly it is a fire, but it shineth not like the
fire of Wood or Coals: But he that will have it to
shine, he must needs add unto it a subtile or fine
Sulphur, that he may extract or allure forth of it a
visible fire.

This fire being defended against the entrance of

the Air, remains occult for many thousands of years, and doth not manifest it self, unless any one make it manifest.

Truly it is an admirable fire, and most fit for the effecting of many incredible things, whereof we have spoken many things sufficient for this time.

XI. An evident demonstration of such a fire lying hid even in the Salt of the Kitchin, and that known to every one.

After that PLATO and many other Philosophers took notice that nothing endowed with life did consist without Salt, and that dead Carcasses themselves were preserved for a long time from putrefaction by the virtues of the same. They thought and wrote that a certain divine thing lay hid in it. But after what sort this divine and hidden thing is to bemade visible, they have not taught. But without doubt, those most wise Philosophers would by this word shew and denote something of a singular excellency.

Because therefore God himself is a fire, and hath never appeared to his Saints in any other shape but that of fire, and besides also all Salts are generated in the moist bowels of the earth from an Astral fire, and on the contrary, a true fire may by the operation of Art be extracted and rendered palpable and visible out of all Salts, it being that which withoutdoubt laynot hid unto them, therefore it is also very likely that those Philosophers have not without a cause of great moment written that a certain Divine or fiery Being did secretly lurk in Salt.

But that they have intimated not any thing to be better, or more noble than that fiery and saltish Spirit may be foreseen by an easie conjecture; for if a certain divine thing shall lie hid in Salt as they write, it shall of necessity follow that that divine spark being freed from all its earthly bonds should be far-superiour to all earthly things in beauty, virtues, efficacy and power; and that next to the eternal God himself it should remain the chiefest and most precious Pearl of the World.

But who shall teach us the manner of separating so precious a Pearl out of the common and Kitchin Salt? None but God alone, or some good friend; who can make his friend a partaker of the knowledge received from God?

But since that very few mortals do seek, love, fear and honour God with sincere hearts, but do much rather cleave fast unto the frail and unjust Mammon, and attribute divine honour unto the same; its no wonder that God doth reserve those things to himself, or at least doth sparingly bestow on us those things which he abundantly supplyed the Ancients withall from his own bountifull hand: And moreover the same omnipotent Creator enlightning some fit subject, with a certain spark of nature, grants unto him also so much wit that he knows that by a due silence he is to beware of this wicked dreg or dross of the World. Whence it is no wonder that the light of nature is at this day made known to so few mortals.

But before I treat in many particulars of that precious Pearl of Salt, it seems altogether necessary for me, first to shew the manner and reason of extracting that fire out of Kitchin Salt; the separation whereof can be perfected in no other respect than through the violence of common fire, to wit, when as the Salt being mixt with a certain earthly matter that it cannot flow, is urged in a retort with a most strong fire, that the more pure part of the Salt, which is nothing else but a sharp spirit, may depart into the Receiver joined to the Retort, in which sharp and sweet spirit a most efficacious fire lurketh which in manner following is to be extracted and concentred.

XII. Of the preparation of the fire of Salt.

Take of this acid or sharp spirit of Salt, rectifie it out of a Glass Retort in sand; the flegm will come over first, which was put in the receiving Vessel in the first Distillation to condense or collect the spirits the more commodiously. After that all the flegm is come off, and acid drips begin to come, remove or change your Receiver, and take your spirits therein; continue the distillation so long until all the spirits be come forth, it being indowed with an acid sweetness, is an effecter of very many operations, which doth bring much profit both in Medicine and Alchemy, as is manifest out of diverse of my writings, and especially out of the 2nd part of my Furnaces, and the comfort of Mariners.

In this sweet and sharp spirit like Wine there is an infernal fire hidden, which doth equally like Coals

27

burn up all things put into it, like as the fire of wood and coals doth Vegetables and Animals, and it reduceth all things which common fire doth, by calcining them into ashes, such as are immature metals, tin, lead and the like, which when they are put into it, it burns them up by calcining them into white ashes.

XIII. A Concentrating the rectified Spirit of Salt into a moist and cold Fire.

Every Spirit of Salt consisteth of two things, to wit, Fire and Water, which water the fire doth so firmly co-knit to it self, that it cannot be wholly separated by any distillation or rectifying; but it always adheres to the fire, how often soever it be rectified or distilled; if any one therefore desireth by rectifying to separate them he must of necessity put immature metal like subjects to the Spirit of Salt, the which, by how much the more immature or unripe they are, by so much they render the spirit of salt the purer; such are LAPIS CALAMINARIS, ZINK, and IRON, which by reason of their moist and attracting nature, do draw to them that invisible fire out of the spirit of Salt, as it were that agent whereof (as to their maturity or perfection) they are necessarily destitute, and without which fiery agent, a metallick kind of body is able to attain unto no perfection in the earth.

Such metallick subjects therefore, the spirit of salt, they being put into it, assaulteth, and as much as it can dissolves them. This solution being distilled out of a glass retort by sand, with the more gentle fire, sends forth nothing but a meer and unsavoury phlegm, the fiery essence it self remaining with the mineral in the Retort, the which if it be more and more urged, and the fire more increased, that it may become plainly burning bright, then that mineral cannot longer retain the fire of the salt, but dismisseth it, which descending into the receiving vessel, is condenced into a thick and fiery oil, which is afterwards to be kept in strong and well stopt glasses, because it fumes without intermission, and desires to return into the air, as it were its Chaos from whence it came forth.

This fire is the operator of great effects in Alchemy and Medicine, of which effects very few have known how to discourse. But it hath far different properties and qualities from that which is extracted out of Vitriol, whereof it shall be afterwards treated.

And although through the help of this fire, incredible things may be performed as well by Chymists as Physicans, and other Artificers; yet it is a consuming, destroying, and also a ripening fire; neither hath it the least of the most noble Pearl with it whereof we have made mention above, and the which in this preparation is converted into such a fire.

That Pearl, if it should be extracted or allured out of Salt, in my simple opinion it were to be extracted not by the benefit of the fire, but through the endeavour of metallick and attracting subjects.

But although I do not profess my self to be so skillfull a Master, and do not arrogate to my self the knowledge of so precious a pearl, yet I cannot but bewray that small little spark of Nature which God hath granted unto me, that so every one may have a clear knowledge and sight of what admirable mysteries Salt doth hide in its own vile body.

XIV. The manner whereby that most precious Pearl of Salt may at least wise in some respect be rendred conspicuous or apparent.

Even as I have admonished in my foregoing writings, that the powers, colours, and virtues of all Vegetables, Animals, and Mm.erals are found concentred in Fire and Salt, so also I now affirm and assert the same thing that by Salt through the benefit of Fire, all Vegetables, Animals, and Metals, may in their own species, nature and properties be increased and propagated into an infinity. So that we have the seeds of them.

For example sake, I prepare Kitchin Salt by the fire, that its tartness being lost, it puts on the nature of an Alcali or LIXIVIAL Salt, I mix some parts thereof with some barren earth, or with naked sand, the which I moisten with water, in these I sow the seeds of vegetables, that they may be nourished by that Salt and may grow, which in thus growing do obtain their own proper figures, virtues and colours, they appear green, yellow and red, sky coloured, purple coloured, and white, & etc. and have a sweet, sour, sharp, bitter, savour, even as God hath bestowed on every particular kind its own proper nature, which operation proceeds from this one only Salt, and the fiery beams of the Sun

30

being tempered with air.

When therefore Beasts are fed with these Herbs growing, and receiving nourishment from the Salt, they are of necessity also nourished and increased by the same; even as also the same Herbs growing from the same Salt do supply nourishment and increase themselves.

But if any could obtain the true seed of Gold, and increase that seed by the help of Salt and Fire; he might (without doubt) obtain great plenty of Gold, but God will not have it that the tail of the Goat should be as long as the Cows, the which being lifted up with too much pride, would strike out her own eyes with her too long tail.

If therefore all things and Gold it self, as also Silver, Pearls, and precious Stones, are after an invisible and occult manner hidden in Salt, and may by the help of Art and nature be rendred palpable and visible; why also might it not come to pass that the most excellent Medicine and most precious Pearl of the wise men might be allured forth out of the same Salt? Truly common Pearls are bred out of Salt waters, wherein if the first matter of Pearls were not, after what manner or sort should they bewray themselves out of the same? Therefore that it may evidently be made manifest, that by the operation of art, also Pearls may be extracted out of Salt, which do far excell those Pearls, which by fishing are drawn out of the depth of the Sea, in beauty, virtue, efficacy and excellency: I will prescribe as much indeed as hath been granted unto me, for demonstrating the possibility of the thing, a certain manner whereby every one shall be able to take to him a firm and sure foundation of weighing or considering of the matter more exactly.

XV. An operation of alluring forth a Philosophical Pearl out of Salt.

Dissolve thou in common water, as much of common Salt as thou wilt, by how much the greater plenty thou shalt take, by so much the more thou shalt obtain.

In like manner dissolve in AQUA FORTIS one or two Ounces of Silver, pour this solution of LUNE on the dissolved Salt, and stir both the dissolutions up and down divers times, that it may become white and like unto Milk. For Silver cannot well indure the Salt, but departing from it is precipitated to the bottom, and there resides, in the form of a snow-like Powder, which by the effusion or pouring off the water is to be separated and dried.

This silver powder hath extracted a spiritual and philosophical gold, or the said precious Pearl out of the Salt water. Because DIANA hath known no less how to fish Pearls in the Salt Sea, than to hunt wild Beasts in the green Woods: But that Pearl is made corporeal and visible in mariner following.

XVI. How the Pearl being attained is made visible.

It is to be noted that that silver powder being thus by it self, and without an admixture of other fixed Salts, doth very hardly by fusion return into its former form of silver, but that it flows like Salt, and pierceth any vessel whatsoever, yea doth depart into a smoke. For the spirits of the Salt do render the silver so fluid and volatile, that it is made altogether mercurial; and therefore its more tender and noble part may be separated from its more gross part by distillation, if this could be done by glassen, or earthen, or metallick vessels.

When this mercury of LUNE is melted in an open crucible, it vanisheth into smoke. It being put into a Glass Retort, refuseth to yield to the fire, the which being too much increased makes the glass to melt, and destroys the glass together with the silver. If earthen vessels be used, the same mercury pierceth the same unhurt like oiled Leather, when it departs, the Salts also depart into smoke, and do leave little grains of silver adhering to the vessel, whereof in this respect there is made a loss, which renders the sublimation void.

Of Iron vessels also here is no use, because of the Salts that are admixed with the silver rising up against the Iron, they dismiss the silver reduced to its ancient body, and besides a little spirit of salt they send forth nothing, so that no separation is made, but the pure and impure do remain co-mixt.

For the sake of avoiding those discommodities I have tried many ways and manners in vain, and at length I took notice, that if such a matter be added to the most penetrating mercury of LUNE, which may so hinder its efficacy of solving and co-melting, that it may be changed into a porous lump, that then, through the benefit of fire there might be an easie separation thereof, which without this help doth most difficultly exist.

In the name of the Lord, therefore adjoin thou unto thy fishing Net; that is, unto the mercury of LUNE, such a matter in due weight and measure which admits not of melting, and which suffers not the mercury of LUNE to conflux, or melt together. Such are wooden Coals being reduced into a fine powder, with the which being mixt with the mercury of LUNE, thou shalt fill thy distilling vessel, whether it be earth, or iron, or glass which is the best of all, even unto the half part, and shall set it in the fire, the which is to be gently increased by degrees, until the glass become burning bright, keep the vessel so long in this heat until all the spirits are departed, which ceasing, thou shalt take away the vessel being cold, in which thou wilt find the remainder of the mercury which did not ascend, reduced into a corporeal or imbodied silver, or at least wise such, to which adding a little borax is easily reduced into silver, the which doth contain somewhat of gold; but keep thou that subtile and pure matter which ascended in distillation as a precious treasure, and meditate after what sort, or by what means thou mayst be able to fix this precious Pearl, and convert it into a fusible, or flowable, and piercing stone.

But in what respect, or in what manner this thing is to be done, in very deed I cannot tell, because I am he who have not hitherto had leisure, nor time of perfecting that thing, and therefore I have been willing here to shew only these things which I have seen with my eyes, and handled with my hands.

Another shall be able by his own judgement to make trial, and to see what God will bestow upon him, I have shewn instead of the mercurial Statue or Image, that which shall suffice at present.

XVII. A more eaise manner of obtaining a Philosophical Pearl.

If thou shalt be desirous of obtaining a Philosophical Pearl after a more easie manner, thou must of necessity thus operate.

Unto half a Lotion, i.e. two drains of the mercury of LUNE, add a little of the powder of Coals, and put the conjoined matters into a small glass, the which set in a crucible encompassed with sand unto that height which the matter in the glass it self shall determine, On the mouth of the glass put a small piece of some glass that it may be well covered, and so place thou a less crucible with the upside downward upon that little glass, that its (top) utmost and highest bound being overwhelmed with the said sand may drive away all air from that little glass.

Set that crucible being in this manner co-fitted, and containing the little glass shut up between them in live Coals of Wood, and make them bright burning hot, that that may remain fired for a quarter of an hours space, then let them cool, and thou shalt find a little lifted up by sublimation, the rest being melted by borax, will afford a silver impregnated with gold, yet without gain, the which demonstrateth only in the space of half an hour, what may be done; but what gain may be obtained by this very operation shall hereafter be shown.

Furthermore it is here to be seen how most beautifull a Pearl doth bewray it self, although very little of it come forth, because in this labour no small part thereof flies away into the air, and sheweth only its colours alone in the glass, far more beautifull than gold, silver, and precious stones; if any one shall rightly operate, neither shall there be any Painter who shall express it by imitating and painting.

For this time take what hath been spoken in right and good part, and immediately weigh thou so great a thing the more exactly, pray, labour, seek, and in seeking thou shalt find such things which thou couldest never before have believed.

The Brethren of ignorance, my enemies, will here object against me, and say, that these most elegant colours have drawn their original from the silver; unto those I briefly answer, that they were indeed extracted out of the Salt by the help of the silver, but that they do not (per se) or by themselves pertain to the silver, for if they were of the silver they would also be solved by AQUA FORTIS, the which, since it is not done, they are not silver, but the meer ANIMA or Soul of the Salt. That this thing may be confirmed by a more evident argument, I bring the solution of Saturn or Lead, the which it self also can fish out the same Pearl from Salt, without Silver. If any one shall operate after the same manner which I but now shewed, I also add this,

36

that I am hereafter to teach a way whereby SATURN may be able to fish pure simple gold out of all salts.

Let us now return unto the moist and cold fire of the Philosophers, and see what an admirable fire GOd hath hidden in Salt Peter.

That a most potent fire doth lurk in Salt Peter is not worth our confirming by any argument. That horrible Gunpowder which shakes or rends all things asunder proveth the thing most manifestly, and AQUA FORTIS, which dissolveth and destroyeth all Metals, yet another fire of far more powerfull virtue is hidden in the same salt, which very few have known and beheld, and the which we will here make manifest, for the honour of God, and the profit of all mankind.

XVIII. Of the preparation of the moist and cold fire of Salt Peter.

Take of Potters earth being without sand, and burnt, 2 parts, and 1 part of Salt Peter very well purified, with both these matters being ruduced into powder and well mixed together, fill a glass retort well coated with clay, put it in a Furnace for distillation, and join a Receiver to the Neck of the Retort, into which put as many pints or pounds of water as there were pounds of Salt Peter mixed with the earth, that the Spirits going forth may so much the sooner be condensed into moisture, after thou hast exactly joined and luted thy Receiver to the neck of the Retort, with a due lute (or clay) kindle a fire according to Art by degrees, and the spirit of the Salt Peter (representing a yellow or red mist in going forth) will join it self to the water placed in the Receiver.

All the Spirits being come forth, take off thy Receiver, and separate them from the water, put this sharp spirit of Niter into some strong glass; it being by distillation freed from its superfluous phlegm and rectified, is applied unto Medicinal And Chymical uses; concerning the operations and virtues whereof there is mention made in the second part of my Furnaces, and in the Dispensatory of SCHRODERUS. Moreover, the manner of extracting and concentrating a fire of this spiritus thus.

Pour this spirit of Niter on the powder of LAP. CALAMINARIS or ZINK reduced into small little grains that it may dissolve as much as it can; and when it will dissolve no more in the Cold, place the glass in hot sand that it may dissolve more of the matter, filtre the solution and by sand draw off all the phlegm in a glass retort; the phlegm being all come off, change thy Receiver, and increase thy fire and drive out a fiery oil, which oil thou shalt keep well stopt, because it uncessaritly fuming would wholly vanish away in the Air.

This fiery smoke of Salt Peter, as also that of Vitriol, and common Salt, burns up all Herbs, Grass, Leaves and Flowers, and whatsoever it toucheth, just as if they were burnt with a strong heat of the Sun or Fire.

And this is the preparation of the moist and cold fire of Salt Peter, of the use and efficacious operation whereof in Medicine and Alchemy, it shall be more exactly and fully treated on in the following Chapters.

XIX. Of the moist fire of Allome.

Allome also by the work of Distillation and
Concentration yields an efficacious fire most like to
that of Vitriol, in efficacy and virtues, but the plenty
doth not answer by reason of too much earth wherewith it
abounds, yet if somewhat of the other Salts be added
unto it, rightly and orderly bestows its fire.

XX. Of the moist and cold fire of Sulphur.

Although Sulphur finds not a place in the order of
Salts, because it refuseth the solving in water, yet it
contains a vitriolated salt laid up in it, which doth
not manifest it self before that the more fat substance
thereof shall be withdrawn by inflaming, by the
operation whereof the salt is attenuated or made thin,
and is carried on high by the flame like a sharp smoke,
so that this sharp sulphureous spirit burns all things
which it toucheth, after the manner of all those fires
which are drawn out of salts.

For the attaining this vitriolated and sulphureous
spirit the flame of the sulphur is to be received, in a
certain Alembick made of glass or earth, peculiarly for
this operation, wherein that vitriolated spirit of salt
condenseth it self, and issues forth like a thick fat,
and fiery oil, not unlike to that which is made of
Vitriol, whereof it is treated on in my Furnaces.

All these things do very evidently confirm those
particulars, which I have many years ago committed to

memory; concerning Sulphur and Vitriol, to wit, that Sulphur is the original of all metals, and that no metal at all is digged out of the earth, which hath not either Vitriol or Sulphur, or for the most part both adjoined unto it, for no sulphur is destitute of vitriol, nor vitriol of sulphur, so that both of them do challange the rise or birth of any kind of metals whatsoever unto themselves. And every sulphur is by its own proper agent or vitriolated salt, which it hath in its possession by nature (whereto the central fire of the earth is sri assistant) excocted or boiled up more and more into a metal; neither doth this universal agent or vitriolated salt depart from the fatness, or its patient, until the fatness together with the agent shall depart into a malleable metal, or a metal that undergoes the hammer; Lead, iron and copper, do make this thing manifest, which metals do never appear without vitriol and sulphur, and that for this cause; because they being as yet unripe and imperfect ones, do stand in need of their agent. A less plenty of Sulphur or Vitriol is found with silver, than with Copper.

Gold hath little of Vitriol or Sulphur, yea plainly none at all, if it shall attain to its highest maturity, because it is then found to be pure and malleable, and wants not a further fusion or melting, but by how much the more of Copper, Gold, and Silver have, by so much the more of Vitriol or Sulphur they have, as also require the more time for their ex-coction and perfecting.

From these particulars, it manifestly appears in

what respect metals may in a long time be generated in the bowels of the earth by their first principle, namely Sulphur; and may be ripened to perfection, by its own Salt, or agent, which it hath in its possession.

If nature doth effect this in a long time, why also may it not come to pass, that art should perform the same in a shorter time?

But let these things that have been spoken be sufficient, he that understands not, nor also perceiveth the scope or mark, which I so clearly shew is blind, and doth not admit of a remedy for his blindness.

Truly I judge these few things, (but yet such as shew a most long way with a most shining Torch) to be sufficient concerning the moist and cold fires or minerals, by which the ripening and perfecting of metals, are to be perfected as well by nature in the bowels of the earth, as by art above the earth.

N. B. If therefore a mineral may by the help and impulse of its own vitriolated Salt, wherewith it is endowed, be ripened from its vile form and lowest degree unto a better, and at length unto the best of all, that is, unto the purest gold, it being that which none (that is seasoned but with the least knowledge of natural things) will deny.

Also if such a Sulphur is hidden in any vegetable, which answers to a mineral Sulphur in its nature and properties, why also might it not come to pass, that this same Sulphur might be perfected into mature gold, alike equal to the other? From hence it most evidently appeareth that in any Herb, although the most abject one, which is promo-ted by the Sun unto its maturity, a spark of the immature beams of SOL may be found, which through the operation of art, are to be changed into pure gold. But after what manner such a Sulphur may be extracted out of any Herb or any Wood whatsoever, in all things like to a mineral one, I have long since delivered in my little work concerning the nature of Salts, and in the second part of the MIRACULUM MUNDI, and below I will demonstrate by a much more clear manifestation.

Let us proceed to Animals and Vegetables, and consider whether in these very things, such a ripening fire may be found, and may from thence also be drawn and made visible.

But we must know that no small living creature or small Herb can grow, live, and receive, increase without a certain fiery and Salt Agent; the which although it cannot be believed by any one that is lifted up with pride, and of a stupid brain, yet it in very deed existeth, and can easily be demonstrated by the hand of the Artificer.

XXI. A most powerfull manner of extracting a fire out of any wood, or any Herb whatsoever, and of rendering it palpable and visible.

Fill some glass, stony, or earthen distilling vessel with any dried wood or dried Herb, and distil off the Vinegar or sharp liquor from thence, and separate the Oil from it; and pour that sharp liquor on LAPIS CALAMINARIS, ZINK, or ashes of lead, which matters do dismiss all the unsavoury moisture in distilling, and retain the whole sharpness with themselves, the which being distilled from thence ascends like unto meer fire, it being of great use as well in Medicine as Alchemy, whereof mention shall be made hereafter. But here it is to be noted that this fire extracted by distillation, is only a part of that fire of the wood and Herbs, and that the other part remains in the Coals thereof, which is far more fixed than that which ascended, and is that Sulphur which we spoke of but now, which wholly answers to the nature of a mineral Sulphur, and which may be extracted out of the Coals being solved by SAL MIRABILIS, which shall be taught and manifested in the following Chapter.

For if there were no fire in them, after what sort should they burn and draw out heat? All Coals being converted into ashes, after that their hidden Sulphur hath done its office, the feces of the wood remains like dead ashes, wherein as yet lies hid a certain singular vegetable fire, being altogether of another nature, and wholly contrary to that which ascendeth in Distillation. This water being extracted out of the ashes presents a LIXIVIUM, the which by decoction exhaleth all the moisture, and leaves the rest a fiery Salt, whereof in the second part of my Dispensatory. If it be made hot without fusion, or melting, it becomes the more fiery, so that it being bound to the skin for some hours in the bigness of a pea, it burns a small hole therein as if it had been burnt with a bright burning iron. And therefore Chyrurgeons make use of such fires that they may open unripe Ulcers, or make Issues.

It may be seen by these particulars that in any wood or any Herb, there are also fires of divers kinds, the which also are found in living creatures, they being partly volatile and sharp, and partly fixed, and obtaining the nature of ALCALIES or LIXIVIAL Salts.

Both Salts or Fires, after they are conjoined they lose their fiery nature, and get unto themselves another quality and property, to wit, a middle one, and these two contrary fires become an essential tartarous Salt, and sweet in use, wherein no fire appears, although that fire being turned out and in by art, may be again extracted and made visible.

Concerning these wonderfull changes of nature, and conversions out of one species or particular kind into another, many things are found up and down in my writings. In the first part of the continuation of the miracle of the world, it is manifestly described after what sort a plenty of such fire may be attained out of woods, but the concentrating thereof is here delivered. In general it is here to be noted, that one wood or one Herb doth more abound with such a fire than another. But by how much any wood or Herb is the elder, and by how much the longer the Sun-beams have operated on it, by so much the more of fire is in it, as is manifest from the Vine; which hath received pleritifull Rays of that sort, and therefore excells all other vegetables in the greater and stronger fire, as appears not only by the burning spirit, but also the tartar, or tartarous Salt thereof, which is almost all fire, and yet without

Distillation and Calcination it cannot be manifested.

That therefore it may be brought forth into open view, and be rendered visible, we must make use of the following operation.

XXII. The manner of manifesting the fire of the Vine.

Fill thou a Glass Retort with common Tartar, and distill forth the volatile spirit and oil, the which thou shalt separate after a due mariner. Great virtues are in this oil, whereof I have made mention in the second part of my Furnaces. The spirit is to be rectified in B that the fiery substance only may depart, and the unprofitable flegm remain behind; the rectified spirit is to be poured on the fixed Salt, (residing in the Retort, which must be first calcined by a strong fire and made fiery) and from thence again distilled, that the fixed Salt may retain the rest of the unprofitable flegm, and the spirit attain the greater fiery virtues for the performing of wonderfull effects in medicine, the which my writings do teach.

XXIII. Another manner of extracting or drawing forth a far more stronger fire out of Tartar.

Dissolve thou that Alcalizated Salt from which the spirit was abstracted in rectifying in a little water, that it may become a very sharp LIXIVIUM or Lye; pour one pound of this LIXIVIUM on two pounds of white Tartar in a Gourd, and that being reduced into powder, put on a head which being well luted on with clay, set it in sand and kindle a fire by degrees, if thou shalt rightly work thou shalt obtain a most subtile fire, one drop whereof doth burn the tongue, as if it had been touched with a burning Iron.

How wonderfull things may be effected by this fire, I have already shown in other places of my writings.

XXIV. A manner of drawing forth as yet a more vehement fire out of Tartar.

Take of crude Tartar and the REGULUS of MARS, or the purest metallick part of iron, the SCORIA being separated, equal parts, the which thou shalt mix by beating together, put them in a crucible with a cover so well fenced with clay that it may admit of no air, keep them in a bright burning fire for the space of an hour, then take them away.

From all these particulars it is made known to every one that a vehement fire lies hidden in vegetables readily serving for the effecting of many admirable things in Medicine, Alchemy and other arts, from the declaring whereof the shortness of time and this treatise, commands us at this time to cease. But moreover we must see whether living creatures also are potent in the same fire, and in what respect any one may be made partakers of the same.

XXV. The preparation and Concentration of fire out of Animals.

As the Vine is the most noble of all vegetables, so man also is esteemed by all that are indowed with judgement to be the most noble of all Animals, or living Creatures; the truth whereof the thing it self affirmeth by a plentifull Testimony.

Therefore we pass by all other Animals in silence, and do here shew (by the following manner) the preparation of that fire only that lies hid in Man.

The Ancient Philosophers have called the great World, MACROCOSMUS, and man as it were the lesser World, MICROCOSMUS, and a comparison being made, they have determined that what things are found in the greater World, the same are to be found in the lesser World, that is in Man.

From whence also they unanimously believed, and also committed to memory, that as well the life of the greater, as of the lesser World, doth consist in a saline and saltish spirit, and that this spirit doth bear rule in one place more, in another less. Neither is there any one also, who will or can deny, that the whole earth is filled with Salt as it were its Balsam; and that minerals are alike equally bred thereby in the very bowels of the earth, as vegetables are in the Superficies.

Yet notwithstanding the Salt of the great World is no where more plentifully found than in water, or in the Seas; the which as it is a thing most known, it needs no confirmation. The same thing is to be understood concerning the little world, viz. Man, and although the whole body in all its parts abound with their true Balsam, yet a greater plenty of this Salt and Balsamick spirit, is found in his flesh than in his bones, a greater plenty likewise in his blood, than in his flesh, but the greatest plenty in his Bladder, or in the Salt Sea of the lesser world, the which is hidden to none, but it is the custom not to seek necessary things in remote places, but in places nigh where they are most easie to be found.

Hence because a more plentifull Salt is no where found in man

XXVI. The operation of preparing a fire out of man's Urine.

I have at large delivered this operation in the second part of my Furnaces, whither I refer the Reader; where he shall not only find a manifold composing of this fire, but also its various use in Medicine.

But although it be needless to describe that operation there repeated, yet it seems meet to me (for a more evident declaration's sake) here to adjoin some admonitions which concern it.

XXVII. Observations which concern the preparation of an Animal Fire.

Such a fire is for the most part drawn forth out of man's Urine being purified by it self for the space of some weeks, and is by rectifying converted into a moist and fiery essence as the second part of my Furnaces sheweth; I have there taught a more easy manner of drawing forth the same fire out of SAL ARMONIACK, which is prepared out of Urine, and by the addition of a strong LIXIVIUM it is distilled and rectified,

48

I have also taught the manner of preparing the same fire out of SAL ARMONIACK by the Addition of LAPIS CALAMINARIS, by distilling it through a Retort.

Spirits rightly prepared after these manners are equally profitable in Medicine, Alchemy, and other arts; because they are those which being well made are all of them good, after what manner soever they may be prepared.

But although these volatile animal fires do readily serve for the performing of famous and notable things (and the fixed Salt of Urine itself, may by Distillation and rectification be concentrated into another kind of Fire) yet they are at a far distance from that true Philosophical fire which the Ancients have hidden with so great care and diligence, because that in these preparations the best and chiefest part of the fire flies away and is lost. But this I say, that these fiery Spirits of Urine being concentrated even as I have taught them to be, are indeed able to effect all those things which I have attributed to them, and shall as yet attribute. But indeed they do not coagulate the concentrated fire of the Vine, which coagulation is not the least key for the composing of an universal Medicine.

For when the Spirit of Urine attains this nature, that by coagulating the most subtile Spirit of Wine, (when poured on it) into a Salt, this Salt extracts the soul of Gold duly prepared; the which also, if it be changed by it self, and converted into a dry and sweet Salt, and be fixed, possesseth the virtues of a

Medicament of a most famous and great use in Medicine.

Every one that is illustrated, but even with the least light of nature, shall be able by an easy business, to smell out what may hiddenly lurk under this Salt.

From the most pure Vine is the substance of the Spirit of Wine, which strengthens the heart of man beyond all other things, as also his brain, and other members.
The Spirit of Urine is the purest and most subtile Mercurial Animal Salt, not having its like in penetrating, opening, and resolving.

This subtile Mercurial, Animal and piercing fire therefore, being joined to the most pure vegetable, that is, the Spirit of Wine, that it may be changed together with it into a dry Medicine, any one shall be able by an easy conjecture to forsee what it will effect in Medicine.

But that I may make manifest the errour, and demonstrate the cause wherefore a Spirit of Urine is so seldom prepared, which will coagulate the Spirit of Wine into a Salt, I admonish that a respect be had by every one of the following particulars.

For first it is to be taken notice of, that the most subtile part only of the Spirit of Urine, and not the more gross part, is fit for the coagulation of the Spirit of Wine. If therefore in the preparation of the same, the most subtile part shall be lost, through the negligence or ignorance of the operator, it can in no wise be brought to pass, that the more gross and dreggy part should cause that coagulation.

But that most subtile Spirit doth not only vanish away in distillation through an insufficiency of the Luteirig not being good, but also a great part of the same is lost before distilling, to wit, when the Urine being successively gathered, is constrained to stand and wait too long, so that the Spirit by little and little exhaleth and departs into the Air, especially when it is gathered together in the Summer or Winter time, for that fire not being patient of any extreme, is expelled by a little heat or cold, and therefore the fittest times for collecting the same are the Months called MARCH and MAY, or SEPTEMBER and OCTOBER, in which Months the Air is temperate, neither too hot, nor too cold, those Months therefore are the fittest for collecting and extracting of an Animal fire out of it.

Furthermore, CALX-VIVE or UNSLAKED LIME is to be added to the Urine (when putrified) and distilled, that the insipid water may be so much the more easily or readily separated from the volatile fire, the which is not done if it be distilled per Se.

I would not pass by these few things in silence for the sake of the Reader, and of him that is studious of good Medicines: But after what mamier Metals may be amended by this Animal fire is not here shewn, but God willing shall by and by in the following Chapters. But we put an end to the preparation and concentration of Animal and Vegetable fires, with these sayings, whose admirable virtues and faculties in medicines, Alchemy and other profitable Arts, shall here be manifested in order, as much as time will permit.

Look I pray you on the Elementary Sun, as also on the fire of woods, and the virtues of light, and the virtue of both, the which all creatures, and especially mankind it self, is constrained to make use of for their own safety; could even the least grass betray itself? Or any small worm be bred and live without the Sun? Could any workmanship or artifice be exercised without the help of common fire? The which, if it were not, we should be constrained to eat unboiled Herbs, and raw Flesh like wild Beasts; yea, the whole conversation and negotiation or traffique among men should be wholly taken away, if earthly fire and light should be wanting unto us.

If there were some one man only in some whole City or Province, or in a whole Kingdom, who alone could make others partakers of fire and light, would there not be made the greatest concourse of all men unto him? But because it is known to every one, and every one hath known by an easy manner, how to strike it out of flints, it is had in no esteem, for it is customary not to esteem those things which are made common, although they are precious. The same thing hath happened to the fire, the which although it ought to be made of greatest account, yet it is reckoned of no worth because it is common and vulgar.

But even as the common fire,and that known to every one, doth by very many most profitable operations bring much good to mortals, who can least of all want the use thereof; so also I affirm that those artificial and hidden fires are to be very much accounted of, because a Physician can hardly be without them, for the preparations of efficacious Medicines, and a Chymist can never want for the transmutation of the more base metals into better, either of them without the aid of those fires shall perform nothing of any great moment in Chymical Labours.

He that works and is ignorant of such fires, what will he effect in metallick operations? He being conversant in cold and darkness is afflicted with the same difficulty, as a certain brewer or baker is, who wants wood in the winter season, or who is not able to use water, it being congealed into ice, the one he cannot bake although he hath the best meal, and the other brew drink although he have abundance of the best malt.

So also goes the matter with Alchemical Affairs, the want whereof causeth that we handle not the most noble Alchemy with any profit, but rather receive loss from the same, daily experience being witness, that 100 are wont to be sooner undone than that it happens to any one man to get himself riches thereby. The blame of which discommodity is not to be transferred on an impossibility of the Art, but rather to be imputed to the want of those moist, cold, and ripening fires extracted out of Salts, the which after what sort they ought to be used for the amendment of metals, as also for medicine and other arts, shall be taught partly in this, and partly in the other Centuries.

XXVIII. The general use of our concentrated fiery and ripening Spirits, extracted out of Salts, in the amendment and converting of metals into more noble ones; also the preparation of many excellent medicaments, and the increase and amendment of many other arts, are briefly here demonstrated; the which, God granting, shall more largely be declared in their particular use.

That I may discourse in few words whether imperfect metals may by the operation of the more common and gross Salts, and of the fire be broken, destroyed, cleansed, and reduced into a better form, it being that which the fifth part of the prosperity of GERMANY confirmeth by divers experiments.

I affirm that the pure Spirits of Salts, do with a greater efficacy, and far better effect the same, the which, since those simple Spirits are able, better and more easily to perform than gross Salts, why should not also concentrated Spirits after the best and easiest manner of all perform the same thing?

From a like reason the use of Salts shall not be of so great efficacy in the preparation of medicines, and other arts, as the using of common Spirits is; the which, notwithstanding being still for the most part clogged with much phlegm, do of necessity not disclose so great virtues, as those concentred fiery Spirits do which are freed from all phlegm.

The Sun-beams are for an example which do not send forth so great heat, when they are co-mixed with a moist air, as also green and wet woods do not so vehemently burn with heat, as withered and dry ones are wont to do.

Yea if the hot beams of the Sun are concentred in or by some hollow glass increasing the fire, or the fires of Coals by a strong blowing of the Bellows, and are as it were constrained into straights or narrow passages, they effect ten times, yea one hundred times more than those which are not centred together after such a sort. But by how much the more strictly those forces of the beams of the Sun, or of other fires are concentred by so much the greater, stronger, and sharper heat they draw out.

A burning glass of one foot Diameter, only enflames wood; but one of two foot Diameter will melt Tin, Lead, and other metallick matters of that sort, which are easie to be melted, as BISMUTHUM, or the whitest, lightest, and basest kind of Lead, ZINK, the non-splendent metallick dark matter KOBOLTUM, & etc. But if you extend the Diameter to four foot, the Sun-beams taking the stronger increase will melt silver and copper, and will render iron it self so bright burning hot, that it may be wrought with a hammer, as if it had been heated with Coals. This effect is to be ascribed unto the concentring of the Sun-beams by an instrument, and to the constraining of the heat of Coals, by Bellows, or Wind.

The same thing is to be understood concerning our concentred and moist fires, which ought to be compared, not only with the common beams of the Sun, or with the heat of Kitchin fire, but also with those Sun-beams which are concentred by a glass, and with the fiery heat of Coals constrained or forced by windy blasts. Whence they must of necessity be of greater virtues than the common Salts, and watery spirits of them, the which the more quick sighted will sufficiently comprehend and believe. Simple Country People do see this thing with their eyes, and handle it with their hands, as well knowing that the subtile, hot, sweet Spirits of Wine and Ale, (and those procuring strength to the heart) when they are freed from all moisture by Distillation, and concentred by Rectification; effect ten fold more than if they had still remained with their humidites.

That thou mayest understand the thing more clearly, well weigh thou Grapes, Bread-corn, or the Fruits of Trees, which we eat in that substance as the trees bring them forth unto us; and they afford us a nourishment, but not such a one, as their juice being pressed out, and separated from its dreggs, and by fermentation reduced into a clear and sweet drink.

If necessity compell, Bread corn may be used for nourishment as it is, yet not so well as when it is separated from its husks, being changed into meal, and reduced by water into a mass or lump, and Salt and Leaven added, and by Fire concocted or digested into Bread of the best Savour. By the same reason indeed somewhat better than the water it self, but if it be artificially handled, and boiled up into Ale or Beer, the husks are separated from the more pure juice, the which afterwards by fermentation, separates many dreggs from it, and arrives to a more noble nature, yielding a sweeter and better drink. But if the same juice be after that brought by distillation into a greater purity, and concentred together by a narrow compass, (because it is a meer fire) it will exercise far greater virtues, than gross Bread-corn which wants a power of exercising so great virtues.

So also doth it succeed with concentrated Salts, to wit, when the dreggs are separated from them by the help of art, and the more pure parts converted, and concentred into a fiery substance, performing effects of great moment in Alchemy. But that Salts do commonly destroy metals, as well by a moist as a dry way, is known to every Barber, and persons of no reputation. But after what manner metals being destroyed may be reduced into more noble bodies than they were before, there hath been none hitherto (who being skilled in that artifice or craft) that have not hid it with the greatest care. Hence it hath come to pass, that nothing of profit hath been perceived form metallick transmutations, and Alchemy it self hath been made a mock of by the most unskillfull rout of ignorant ones, as if it were most false, and at the farthest distance from truth. That this doubt therefore may be taken away, and the truth it self may be more evidently placed in our view, I have resolved in my mind, by God's assistance, to place before the eyes of the whole world, a true and profitable transmutation of metals, by a clear description, and to assert the certainty of so many writings set forth by such men, by the most true experiments, so that every one that is seasoned but even with a light of small knowledge of the fire, may by an easy business hereafter obtain some profit from them. But I will first treat of common and crude Salts, and then of the simple Spirits of those, and at length of their concentred Spirits and Fires, which we have taught to extract out of them.

But before I attempt to describe and assert this kingly and noble art, I have been led first to shew the cause why some places do occur in the description of the same, wherein words are omitted, and signs or blanks reposed instead of the same.

Indeed this was therefore done, that the art may be concealed from the unworthy, and they in all respects to be driven from the same, and may be made known only to Adeptists, and the Sons of Art.

Besides also that all secrets may not in all places, and without difference be divulged, but that the chief things thereof may be preserved for friends, lest they be trodden under foot, and broken to pieces by the unworthy, but that they may be left to friends as it were a certain secret stroak, and that sri unknown one to others, for to fight successfully.

I therefore earnestly require of every one by a friendly Petition that he be not suddenly angry, if he be not able clearly to perceive, by the sharpness of his wit, all those things which I propose, but rather let him consider that they are not written for him, but for others; by whose capacity they can be perceived. Neither is it altogether necessary that all do know all things, neither also would it be of concernment if friends and enemies attain all those things in their understanding alike, without any difference, which I here openly produce by my descriptions; it is sufficient that some

58

only, and indeed those that are worthy may clearly and knowingly possess the same, and testifie the truth.

XXIX. An infallible practice of changing the more imperfect Metals into more perfect ones by the help of crude Salts.

As I have already a little before, and also in other places of my writings, evidently enough demonstrated that Salts, or the spirits of Salts, are in the earth, or out of the earth a universal Agent, promoting the maturation or ripening of metals: So here I again firmly affirm the same thing, and do say, that by Salts the gross bodies of metals are destroyed, and trans-changed into more noble metals, and that indeed after divers manners, and that more easily or difficultly as any one shall be more or less conversant in Chymical labours.

I will hear God willing make manifest all things, yet not to every one, but to the worthy only, and that indeed after the manner of a Clock or Watch-maker, who taking some Clock or Watch in pieces, do lay up all the parts thereof in some place without any order; the which he that is unskillfull in the art, shall never again compose and reduce into order. But another who before hath handled that art, will by an easie labour again conjoin all those parts, and reduce them into the former body of a Clock or Watch.

All those therefore who have experienced the foregoing labours to be perfected by the fire, shall by these my descriptions easily dispatch or accommodate themselves in future things, not easie to be understood by the rout of ignorant persons which have made no experiments in the fire; who will in vain look into those things which I have written; no otherwise than as if any one being plainly unskillfull in reading and writing, should behold written letters, and knows not what they signifie, or what argument they may contain:

Such a man if he would be angry with the writer, should he not do him much injury, because as being far remote from the fault of that ignorance which hinders him, whereby he cannot read these letters which he had never learned to read.

The same thing must be understood concerning my writings, which are openly published, not for the sake of any one, but only of those who have first learned to understand those kind of writings.

But that I may set upon the thing it self, and may teach the amendment of metals for the better, and shew the very foundation of the whole business, I say, that a true changing of them is attempted in vain, unless they are first destroyed, and wholly slain. A grain of wheat, as Christ himself saith, will never increase or multiply, unless they are first destroyed, and wholly slain; and unless it first putrifies in the earth. If therefore metals ought to be destroyed by Putrefaction, that must needs be done by the help of Salts, according to the truth of the Philosophers Maxim: the corruption of one thing, is the generation of another. The death of one thing, is the life of another. Since therefore metals must die, it must needs be that death be brought on them by enemies, or contrary things, because nothing in natural or artificial things dieth, unless it be slain by its own enemy.

Since metals therefore are to be destroyed, and killed by their enemies; it is of necessity that they are invaded, tortured, and so long vexed by the same, untill the Agent as the stronger part, be overcome by the Patient as the weaker part; that it be slain by it, (or rather the Patient be overcome by the Agent) and be translated into a better nature, in which action the Patient ought not to depart from the Agent, but to be tortured with an un-intermitting torment.

Whosoever seeing his enemy and conceives himself of the weaker force, irideavours as much as in him lies to decline him, by retreating but all occasions of running away and slipping aside being taken away, he is constrained to deliver his life to his enemy, who handles the Patient or suffering party according to his own pleasure, and doth whatsoever he will, therefore after the same manner is the melioration of metals, the which although they should be melted together with.Salts their enemies, yet would they make little account of them, but would separate themselves from the same; so that every part of them being unhurt, would keep its own nature and essence. But if the Salts do take away the occasion of flight from metals, and do inclose them in their Prisons, that they have not way of escaping, but remain, suffer, and die, then they obtain victory over the Salt, and of slain metals are made more pure and better.

This thing is done in the fire by the moist and dry way, of which enough hath been spoken already.

This is the whole and intire art, and there needs no other superfluous teachings; yet he whom these things doth not suffice, let him read the following operations thorough, wherein he shall find truth, and see with his eyes, and handle with his hand; those things which have been heretofore impossible to him, and very many more.

XXX. After what manner Metals may be slain by their enemies and be transmuted into better.

Unto Metals not one but many enemies are adverse; and part of those enemies are enemies to some and friends to others, but the other part is friendly to some, and at enmity with others, For example sake.

Nothing more prosecutes Gold with an hostile hatred than burning Sulphur and sulphureous Salts, such as are Alcalies, and crude tartar; the cause of this hatred is, because Gold is nothing else, but a fixed Sulphur, and therefore it disagreeth by a capital hatred, with every burning Sulphur; Silver and Lead do love every SULPHUR, AND ALL SULPHUREOUS SALTS, SUCH AS ARE VITRIOL, SALT PETER, SALT ARMONIACK, AND THE LIKE, the which they stand in need of for their colour; they have an hatred against KITCHIN SALT, BECAUSE IT IS OF A MERCURIAL NATURE, and therefore not requiring its help, but only desiring a Sulphur and Tincture, COPPER, IRON, AND ARGENT VIVE, or Quicksilver do possess both natures, to wit, A MERCURIAL AND SULPHUREOUS ONE, and for that cause they prosecute all Sulphurs, and any Salts with love.

Tin is an enemy of all Salts, whether they are sulphureous or mercurial ones, when it is slain by Sulphur and Salt, and recalled unto life, it obtains a more pure and thin or fine body, whether of Gold or Silver, according as it shall be handled.

Moreover, if any should desire to obtain as yet a better essence out of better metals, its necessary that he slay them by their enemies, and raise them up again by their friends; by how much the greater and vehement the enemies are whereby metals are slain, by so much the more those metals do suffer, and with so much the more famous and better bodies do there arise.

The whole art therefore consisteth in this, that metals are overwhelmed by their greatest enemies, are slain by them, and after death are separated from them, and that by their best friends, are restored unto a better life.

Thou hast the whole art, neither doth any other thing remain than that thou attempt the matter, and set to thine hand.

For example sake, I will add an operation. Slay a light metallick matter by the sharp Spirits of sulphureous Salts, that it may become a white calx; free this from the Salt Spirits, by water being poured thereon, the which being freed, cannot be reduced into a metallick body by any violence of fire. Likewise slay mercurial metals as are B. by mercurial, Salts their enemies, and change them into white calx's, the which being freed from their saltness WILL BE LIQUID OR

FLOWABLE; mix those calx's, to wit, the mercurial and
sulphurious being slain, put them into a double vessel
of cement, cover the uppermost with a certain plenty of
B. fence well the juncture of the cementing vessel with
clay, set them into a cementing furnace, and at the
beginning administer a gentle fire, that the calx's may
rise up against or assult each other, and the fixed
sulphur may bind the fugitive flowable and mercurial
calx's, D. for although in the cement something would
depart into smoke, yet that is intercepted by E. and
after a certain manner is thus exalted into the degree
of F. Too much fire is not presently to be joined to the
cement or plaistering it self, that some time may be
granted to the matter that is swift of flight, whereby
it may adjoin it self unto the fixed matter, and may
also become fixed and constant with the same for four
hours space, therefore the fire shall be somewhat the
more slack, and afterwards for the space of eight or ten
hours, it shall be kept in a clear bright, burning heat,
that G. may not melt; the said time being ended, the
fire is to be extinguished, and the cementing vessel to
be taken away, in G. a black or brittle body shall be
found containing Silver, the easy separation whereof we
shall afterwards hear.

The calx of both metals being coagulated into a
hard stone, if by grinding it be reduced into powder,
and be put into a furnace fit for this thing, a
metallick body will be attained, being impregnated, not
with a little Gold and Silver, especially if the metals
shall be slain, not by the Spirits of common Salts, but

with gradatory martial waters. In this cement, H. is rendred aureal or golden, and I. is silvery, by one and the same endeavour. The profit also it self is of no small moment, especially if this operation be exercised with the greater quantity, and the bigger instruments always to supply or afford Silver being pregnant with Gold for separation.

XXXI. A brief and compendious manner of extracting and rendering corporeal, a volatile Gold out of coloured Flints, Red Talck, Granates or Red Marble Stone, Sand, White Clay and the like metallick earths.

At the beginning, these mineral or metallick earths are to be made bright burning hot, to be quenched in cold water, and to be broken in a mill, into meal or powder.

After that they are thus broken, thou shalt put them into some Waldenburge, or Cullein Can, and shalt pour so much of AQUA REGIA on them, that they may only be moistened, and let them, together with the Can, be placed in a fire of coals, and incompassed therewith, to be made hot; after that the minerals and AQUA REGIS have waxed well hot together, so much hot water is to be poured on those very minerals as shall be necessary for the extracting of the AQUA REGIA.

Put the minerals thus moistened with the water into great pots, and those made of the best earth, having many little holes in the bottom, on which lay paper for sustaining of the minerals that they may not fall out through the holes, but may dismiss the water only. After the first water is gone forth, other hot water is again to be poured on, and this effusion of water is so long to be continued, until it depart with the very same sweetness as when it was poured on, and no longer offers any sharpness to the taste. So the common and hot water brings away with it the AQUA REGIS, and the AQUA REGIS Gold out of the minerals.

The earthen pots may be placed in a bench bored through with holes, through which their bottoms may pass, that so the water may be received in vessels set under them.

N. B. The minerals may also be put into Barrels or hogsheads having a double bottom, such as are used for the cleansing of Saltpeter, that so water may be so long poured on them, until all the acrimony be extracted by the water.

XXXII. After what manner out of Minerals being extracted, a true Salt-Peter may as yet be gotten with profit.

The minerals being after the said manner freed by extraction, they are to be co-mixed with an equal weight of CALX-VIVE and wood ashes, and cast together into an heap under some open gallery or room, that now and then it may be moistened with Urine, or in want of that with Rain-water, as oft as they shall be dryed.

In this operation the AQUA REGIS, which remained in the minerals, and was not wholly extracted by the hot water by the help of the Urine or Rain-water, charigeth the Salt in the CALX-VIVE into the best Salt-peter, the which may be washed off with Rain-water, and boiled up after the wonted manner.

Therefore after the said minerals have been handled for half or a whole year after the said mariner, and are by rinceing deprived of the Salt it self; they may again be (under an open Gallery or Roof so exposed to the air, that Rain come not at it) collected into an heap and be handled after the former manner, for the supplying (in thier own time) new Salt-peter, the which may be done for many years together. So also from that AQUA REGIS which could not be drawn forth from the minerals, a profit is received.

The cause of this Salt—peter, its being made, is this, because the AQUA FORTIS or AQUA REGIS, or Spirit of NITRE in the same waters, contains as it were the seed of Salt-peter, it obtains that nature, that like an Herb it may take an increase from other Salts, and be multiplied; whence perhaps the old proverb arose, to sow Salt, which thing the ignorant have received with mock, saying after what sort can Salt be sown and multiplied, when it is solved and drawn from Rain-water? But it hath lain hid from those, what kind of Salt it is, and after what manner it is to be sown; the which we have here demonstrated, also the saying of the Ancient Philosophers, asserting that Salt may be sown and multiplied like Vegetables.

As to what pertains to those sharp waters, whereby gold is extracted out of minerals, by what skill they are to be handled as also without loss, yea that they may render that gold with profit; the following operations are to be observed.

XXXIII. A way shewing the extraction of a volatile and fixed Gold out of the Water, from which the Minerals are withdrawn, and the profit which may be received by that Water.

The best way is this, into the solution of gold, or into the water which contairieth gold, pour in the solution of LUNE or SATURN more or less, even as you suspect more or less of gold to be in that water: As for example. Let there be in the water two or three half ounces of gold, dissolve thou therefore about two or three half ounces of silver, or lead, in AQUA FORTIS, and pour this solution into the water containing the gold, be it more or less, mix them well together by shaking or stirring, that the water may obtain the form of milk; after they have settled in quietness, shake or stir them again, and repeat this motion for divers times the space of one hour, and at length suffer all quietly to settle to the bottom. Separate all the clear water from the sediment by pouring it out, and strain the sediment it self through a filtre, that the water may be wholly separated from the silver.

This silver is to be dried, and reduced into its former body, after the manner which shall by and by follow.

N. B. If the Silver or lead had not extracted all the gold, the which may easily happen, yet that gold is not lost, for because sweet water whereby the AQUA REGIS is weakened is present, the which now remains unfit for another use of extracting out of minerals; now by the solution of LUNE or Saturn deprived of their gold, a sharp LIXIVIUM made of wood-ashes, and CALX-VIVE may be poured on the same, with which a little ___[1] is to be added or admixed. For do precipitate or fix all gold in solutions.

After this manner the AQUA REGIS is killed, and every metal which it has yet retaineth, it dismisseth like a yellow powder, whether it be gold alone, or mixed with copper or iron, which powder is to be dried, and reduced after the manner which shall strait-way be taught.

N. B. That the water after the total precipitating of the metals, being exhaled in a Copper Kettle unto a thin skin, and exposed in peculiar vessels unto the cold, it will afford thee a beautifull Salt-peter, concreted or grown together into drops or Ice—acles, whereof thou inayst again make an AQUA FORTIS, to be again made use of for the like operations.

He that shall rightly operate shall get so much Salt-Peter as will recompence the charges of the AQUA FORTIS, and AQUA REGIS: So that he shall extract his gold without costs. For five or six pounds of AQUA FORTIS, wherein two or three pounds of Salt is dissolved, and the which hath at length been precipitated by a sharp LIXIVIUM prepared with CALX-VIVE, doth render ten pound of Salt Peter, the which doth answer the price of five pound of AQUA FORTIS, and

---

[1] Blank spaces such as this are to be found in various places in Glaubers writings, as it were hidden. D.H.

this is the manner of extracting gold out of minerals without costs.

XXXIV. Another and better manner of extracting gold by AQUA REGIS.

Take of ___^{2} by torrifying ___ made into ashes, pour the extraction into an iron Pot, and stir the Calx with an iron SPATULA while it boileth. All the sharp spirits do stick fast to the the phlegm alone vanisheth by exhalation. When therefore the spirits wholly concentred with the and are dried, they are to be put into a close Tigil or Crucible upon Coals in a secret Crucible or melting Pot, then the fire expels the concentred spirits into a receiving vessel; the which spirits may be used for a new extraction. A fugacious gold mixt with iron, remaineth with the which Calx being reduced in a Furnace fit for those operations, which the GERMANS do call STICHOFEN, draws out a lead mixt with gold, the which being expelled by a Cupel enricheth the operators with the best gold and Silver.

N. B. But if such lead should not contain so much of gold and silver, as that it should deserve a separation by a Cupel, that is again to be mix-t with and to be reduced into ashes, and the operation is so long to be repeated, until the lead being rich enough in gold, may deserve that separation.

The separation is also to be perfected with the Bellows, lest so great a plenty of lead shoul.d be

---

^{2} This is another of Glauber's blank spaces, verified in the original printed book. -pnw melted out of the Tests, which operation requires much fire, yet the lead may be collected or conjoined in the Test without a wastefull melting, as shall be taught hereafter.

XXV. An easie making or composing of AQUA REGIS for extracting of minerals.

Because a plenteous quantity of spirit of Salt is easily prepared, the Salt-peter is only to be dissolved therein, and with that solution minerals are to be extracted. For the Salt Peter strengthens the spirit of Salt, that it can so much the more easier set upon and snatch to it the tender gold in those minerals.

The same spirit of Salt, may also without any rectifying be administred for this operation, to wit, such as ascends in the first Distillation.

XXXVI. Another as yet more easier way of preparing AQUA REGIS for extraction.

Because silver doth always in this operation bewray its being impregnated with gold, which is to be separated by AQUA FORTIS, the solution of silver is also fitly used to extract after this manner.

Pour the said solution into AQUA REGIS which hath extracted gold, that the silver may attract the greatest part thereof to it self. But the same AQUA REGIS may again extract other gold, and be attracted by the solution of silver.

But if there should be no solution of silver in readiness, the gold extracted is concentred with and the operation is perfected by the means or after the manner abovesaid, by driving the spirit out of the and by reducing it in a Furnace, called by the GERMANS STICHOFEN, as was said before. N.B. AQUA FORTIS being dissolved therein, or poured on or into AQUA REGIS, or the nitrous spirit of Salt, it adds an increase and strength to the AQUA REGIS, because AQUA FORTIS doth corroborate the spirit of Salt better than Salt—Peter.

XXXVII. How the Calx of silver, which hath fished out gold by AQUA REGIS, is to be recovered.

When the solution of silver is poured into AQUA REGIS, and the chiefest part of the gold is extracted; rest is so long granted unto it, that the Calx of the silver may settle to the bottom, and afterwards the AQUA REGIS by pouring it forth is separated, cleared from the Calx of the silver, again to be used for a new extraction; unless perhaps as much of iron had been admixed with it, in which case the white Calx of the silver is put into some Cloath laidin an earthen or glass Tonnel, and hot water is to be poured on it, to take away with it the AQUA REGIS, which is left in the Calx of the silver. The remaining water is to be pressed out of the Towel or Cloath, and the Calx dried, and reduced in the secret Crucible, or is made use of in the concentring of into silver or gold, viz, gold and silver

XXXVIII. After what manner precipitated silver is to be reduced without a loss of its weight.

Seeing the greatest fugacity is procured unto this Calx of silver, so that its formerbody cannot be restored unto it in common Crucibles without great loss, this discommodity cannot after a more convenient manner be prevented than by that which follows.

Mix thou an equal weight of ___ with this volatile Calx, and cast it into a close bright burning Crucible, that is narrow above, and broad beneath; the which after thou hast covered with a Cover, and well fenced with the lute of Wisdom, thou shall melt the matter together, nothing whereof shall depart into smoke, neither shall so much as the least of it pierce through the Crucible, and all the silver which the gold received is by this means attained without any loss.

This silver thou shalt by fusion reduce into grains, and shalt separate the gold from the same in AQUA FORTIS. And thou shalt again apply the silver thus reduced unto a new labour, in which labour thou proceeding without intermission shalt have a continual separatory operation of gold and silver; and this labour thou niayest exercise with great profit in all places.

XXXIX. Another manner of reducing a fugacious or volatile silver, with greater profit.

Place thou at the Stern of this little golden Ship, a little fish whose name is REMORA, that it may be spoiled of its swiftness and may be at a stand, cast this silver little Ship with the little fish REMORA, fitting at its Stern, into a close and square Tigil or Crucible, that by fusion they depart into one body. In

73

this fusion not only all the silver is returned without any loss into its former body, but also is by the white FINNS of the little fish, augmented with a certain increase of its weight, and becomes more golden; so that by this additament more of better silver is gotten, than if by the addition of other things it had been restored to its former body.

What other profits any one may be able to obtain through the help of this volatile silver, we will God willing hereafter teach.

These are the things which I at this time have been willing to teach, concerning the extraction of a volatile gold out of stones, and the more poor minerals, as also of the extending or bringing forward silver by successive degrees into gold; of which matter more things shall be spoken in other places.

XL. An operation, teaching to extract Stones and Minerals, or Mines that are poor in Silver, and Copper by a moist way.

These matters being made bright burning hot, are to be quenched with water, then moistened and extracted with AQUA FORTIS; after the same manner as was taught above concerning the minerals of gold, and no difference is here met with but in the waters extracting, since gold is extracted with AQUA REGIS, and silver with AQUA FORTIS.

If the mineral or mines of gold and silver are at once in readiness, the gold is extracted by AQUA REGIS, and the silver by AQUA FORTIS, and the solutions are to be united, in which dissolving, the silver being precipitated by the AQUA REGIS, doth also snatch with it the gold from the AQUA REGIS; and although copper shall be present with the mine of silver, and it be extracted together with the silver by AQUA FORTIS, yet it is no impediment to the operation, for the silver and gold do sink to the bottom, and the copper is retained by the AQUA REGIS to be afterwards administred for a new operation, and that indeed as often as any one shall be willing.

The copper is recovered from the AQUA REGIS by thin plates of iron being put therein, which operation makes the AQUA REGIS red, and wholly unfit for the like labours.

Therefore the iron being then spiritual, promotes something out of the lead unto the degree of gold, and so the AQUA REGIS being thus often used, it is again rendred profitable.

XLI. A more easy manner as yet by far, of plentifully extracting Gold and Silver out of poor mines, as Sand, White-Clay, and other the like minerals, by fire without fusion.

The mine or mineral ARGILLA, or white-clay, & etc. containing a volatile and fixed gold, beingroasted or calcined, and broken in pieces in a mill, fill thou a glass gourd therewith fenced with clay, or made of the best earth, half full, and pour so much of the following MENSTRUUM on that matter, as that it may be well moistened; but as soon as that MENSTRUUM is poured thereon, it presently begins to give a smoak, wherefore it is altogether necessary, that thou presently put a head on the gourd or body, which is to be set in sand, and all the moisture separated by distilling, and that while the distillation is performing the gold may be dissolved, but the solving matter it self is to be collected in a receiver by it self, the which hath the virtues of AQUA REGIS, and may be again applyed for use, as shall by and by be shewn.

After that all the humidity is come forth, take the gourd (being cold) out of the sand, and pour some water on the matter that it may become soft, and that a Salt may be extracted from it, wherein the Gold lurketh, which was contained in the mine; coagulate the LIXIVIUM being full of Gold, into a Red Salt, the which by adding Litharge, is to be melted in such Crucibles which are not broken.

The Litharge draws the Gold unto it out of the Salt, which is to be separated from the Lead, after that manner which shall be shewn in the following Chapters or Treatises.

XLII. The preparation of a Water necessary for the extracting of Gold.

Take of this water of small charges, which thou shalt prepare plentifully without trouble, pour upon mines, and again separate it by distillation, to be again used in new labours, that there may be no need to prepare it again anew, because this doth not only always remain effectual, but also is increased in every operation: So that thou mayest be able to extract mines and minerals AD INFINITUM, if so be thou shalt prepare but one pound or pint at the first.

N. B. By this means all Gold how little soever it be, is plentifully extracted out of flints, sand, and any other minerals, without any cost excepting fire.

XLIII. Another water for extracting silver.

Take this water extracteth silver out of the poor mines of silver, sand, and stones; the operation of the same, is like the former one, and its increase is like the increase of the former water, so that after this manner silver may be plentifully extracted out of poor minerals, and no other cost is required besides fire.

N. B. Instead of ___ may be taken since it performs the same thing in extracting, which the other performeth.

XLIV. Another easie manner of plentifully extracting gold and silver out of poor minerals, as being of little or no cost.

Mix thou the mine or minerals with the requisite waters, fill with the same of good earth, set them near each other in a great put it in and distill the spirits, which pay all the charges, and which supply gold and silver without any costs, the which is to be received by lead.

XLV. Another more easie manner of extracting gold and silver out of minerals.

Mix the mine or mineral with the requisite waters, and moisten it by degrees, cast the whole into the spirits depart into a receiving vessel, and in the time of distillation, the gold and silver are dissolved by that dissolvant, the which being extracted, remain with or among and are rinced by water out of the mineral, so that they are attained without costs, and the spirits being collected in the receiving vessel, do recompence all charges.

XLVI. An easie operation of plentifull extracting gold and silver, out of far white Clay or Potters-earth.

Although gold and silver be extracted out of minerals by moist waters, by a troublesome operation as we have taught in the beginning, yet such an extraction brings no small profit, because they may be freed from those waters by precipitation; and those very waters do readily serve for the making or preparing of Salt Peter. With a fat Argilla or white Clay, the matter goes otherwise, because the spirit hath crept into the far earth, and scarce a half part is received, unless the earth be first deprived of its fatness by making burning bright, it being that which doth also require its own peculiar labour.

Seeing that therefore in all places of the world, wheresoever earth is found, such fat white Clays containing gold and silver do plentifully offer themselves, and none hath been hitherto found, who hath indeavoured to extract and bring them to use, especially

78

whenas they contain but little of gold or silver, not able to defray the charges of the lead bestowed.

I could not but open a very easie way of performing that thing with no small profit, the which is perfected by the following labour.

Mix such an earth containing gold or silver, with or of that greater one, and cast it by degrees into my first or second Distillatory Furnaces, and draw forth the spirits by distilling, that the solved gold or silver may remain in From that which is broken by a Mill or grinding is washed off with hot water, and is reduced by lead as is shewed before.

XLVII. After what manner by the help of art, gold may be easily and plenteously extracted from the sand of GRANATES, AGATHS, SAPPHIRES, AND RUBIES, and other stony Mineral earths, which do neither admit of fusion, nor Lead, nor sharp Waters.

It is certain that all GRANATES or Marbles, the red, black, ashcoloured, duskish, also of whatsoever colour they are, or wheresoever they are found, whether in Brooks or Rivers, or in Sand, or fat Earth, or in high Rocks, do always contain much of Gold, but that by reason of their glassy nature, they cannot be tamed by AQUA FORTIS, and by reason of their most difficult fusion cannot be wrought by Saturn or Lead; for that cause they have made none partaker of their gold, and they have been neglected as unprofitable earths, whenas notwithstanding they being handled by little labour, they are able to afford much gold.

Some one therefore may ask, because they cannot be subdued either by lead or AQUA FORTIS, what course must be taken, that gold may be extracted out of their bowels? I answer and also by a fusiblethey may be so overcome, that the gold which they have may be withdrawn from them with profit, for because they can bear the fire, they are and perform by a double labour, one through their sharpness, and another by reason of a strong fire, so that nothing is safe from so great forces, but all things are constrained to yield unto so great violence. Hence they do willingly afterwards enter the Salt of Lead, and in time of melting do readily draw out their gold.

XLVIII. A most firm demonstration, that sharp Waters and Salts after the manner of the ways hitherto described, do draw forth Gold and Silver as it were without cost, out of Mines or Minerals containing Gold and Silver, than sumptuous or costly melting Fires.

It is not unknown to every one that is seasoned but even with a mean knowledge of gold-bearing Mines, that without the addition of Lead or other flowable things, Gold can in no wise be melted out of them, whenas therefore there is little Gold in those, and nevertheless much of Lead or other matters is required to be added for an easie fusion or melting sake, who shall be fit for the undergoing of so great costs? Hence it hath come to pass that such poor minerals could bring nothing of profit in common.

It being also granted, that there is so much of

Gold and Silver in Minerals, that they will defray the charges of fire and fusion, and afford Gold and Silver with profit, yet such operations are not comparable to my inventions using Waters and Salts, no more than as Water to Wine, or Night to Day.

For first, fusion or melting far exceeds in its costs the extraction which is perfected by Salt and Waters. And then it can never be brought to pass that fusion should expell all the Gold and Silver out of Mines and Minerals, but leaves some thereof in the drosses, next also there is made a loss of all the volatile Gold and Silver, which are cast forth by a most strong fire, and driven away into the air; on the contrary, of the more ample and wealthy profits, which the extraction of Gold and Silver out of Minerals bringeth, this is not the least, that not onely all the Gold and Silver which is fixed, is drawn out without any detriment or loss, but also those two metals being as yet volatile are extracted at once and made fixt and constant in the fire, whenas notwithstanding by the vehemence of a melting fire, they are wholly dispersed and reduced into nothing. For the Waters of Salts do fix the volatile fugative spirits of Gold and Silver, the which by the vehement blasts of bellows, are rendred as yet far more volatile, so that by the help of fusion scarce half the Gold is gotten which the extraction by the Waters of Salt affordeth.

Add to this, that not only all the Silver and Gold, as well the volatile as the fixt is gotten together without any loss, but also the charges of Coals

81

are far less, and one operatour may perform as much of this extraction of Gold and Silver, as three in the labour of fusion or melting.

From hence it is easily to be seen, how much profit and fruit may be gained by this my most excellent invention in all places of GERMANY. For this extraction is not only for extracting Gold and Silver as well out of rich or poor Mines, but also snatcheth out all the Gold and Silver, in Flints tinged with any colour, wherewith all Brooks, Rivers, and Fields are filled.

Now follow some particular transmutations of the imperfect Metals into more perfect ones, by common fire, and Crude Salt, and by the not common concentred moist fire of Salt.

XLIX. A fundamental and evident demonstration, that a true transmutation, or trans-changing of Metals may be exercised in all places of the earth, wheresoever Men may dwell, yea in the least Cottages of Country folks, and indeed by the same matters and vessels which are found in them.

It is well known, that even the poorest and vilest or meanest Country can want neither fire, nor salt, nor earthen vessels, necessary for the boiling of Meats. Therefore being furnished with Salt, a piece of Copper of some old Kettle shall easily supply him, the which it shall be free for him, by fire and salt to transchange into a better and more noble body, but because Man hath a far better and commodious Salt for the transmutation of Metals in his possession, which excells common Salt in its goodness, he may of right and worthily make use of his own proper Salt before a strange and foreign one, and that after this manner.

Boil thy own, or the Urine of another man, to the consistancy of Honey, in which decoction, all the unprofitable moisture of the Urine departs by exhalation, and the Salt thereof remains in the Kettle or earthen Pot; admix thou with this condensed Urine, so much of CALX-VIVE, or the ashes of burnt Woods, that it may come into a thick or gross lump. But thin plates of Copper cut in pieces, and purged by making them bright burning hot ought to be in readiness, and also sri earthen Pot having its Cover, wherein let that mass of Urine, and CALX-VIVE, together with the thin plates of Copper be put; when thou hast all these things in a readiness, mix the Urine as abovesaid with the CALX or slack-lime or ashes, and fill thy Pot to the height of three or four fingers, upon which put some of the plates of Copper, and so STRATUM SUPER-STRATUM till the Pot is full, then cover it with its Cover, which thou shalt well fence with Lute made of Meal, Water, and Paper, that not any vapour at all may come forth. For as soon as the CALX-VIVE is mixed with the condensed or co—thickened Urine, the spirit of Urine begins to operate by its own proper efficacy, and therefore it is necessary that the Cover be straightway laid on the Pot, and be fenced with a due Lute or Clay; the Pot being thus filled and covered, set it aside in some certain place for the space of half a year, in which time the spirit of Urine being stirred up by the CALX-VIVE, displays its virtues on the Copper, and exalts it into a higher degree, as that it is tinged of a skie and green colour mixed, and is rendred fit for the Painters Art; one pound whereof is more worth than two or three pounds of Copper, one pound whereof doth for the most part render one pound with four or five ounces of this colour, and so it affords a profitable transmutation of Copper.

L. After what manner, out of this partly Green, partly Skie-colour of Copper, Gold and Silver is to be separated.

He that desires to separate Gold and Silver out of this colour, whereof no great masses are attained, but only and alone whereby it is demonstrated that the thing may be done, he must use the following operations.

First, he must take good heed in taking the colour out of the earthen Pot, lest any thing of the mass of the Urine, and CALX be mixed with the beautifull colour, and render it impure.

The thin plates being taken out, they are to be often crookedly bowed, and moved upwards and downwards, that the colour may fall off from the plates like scales. The remaining Copper which is not yet turned into colour, is reserved for a new labour; to be repeated after the same manner. Wine-vinegar is poured on the colour, or the sharp water of Tartar extracted after the Distillation of adust Wine from its Lees, with water by decoction, and it is to be so long boiled in a Copper Kettle, or an earthen Pot glazed, till all the colour shall be solved; the solution being cast into a Filtre, the Vinegar or Water of Tartar only passeth through, and a reddish powder is left in the Filtre, the which being melted with Lead in a Cupel, after the exhalation of the Lead, it leaves a grain of Gold. For the spirit of Urine hath ripened somewhat of the Copper into Gold, which the Vinegar or Water of Tartar did not dissolve but left, attracting only the Copper by solution. And because somewhat of Silver is ripened in the Copper through this same operation, and is dissolved with the Copper, by the Vinegar or Water of Tartar, something of common Salt is to be added to the Water of Tartar or Vinegar, that the Silver may not be solved by the same, but may remain with the CALX of Gold, and may be retained with the same.

N. B. Pure and clear Vinegar ought to be taken for this labour, if you would preserve the colour unhurt, for after some part of the Vinegar is evaporated away, the green colour of the Copper grows together into fair green small stones, one pound whereof is more to be esteemed than five or six pounds of Verdigrease, which is sold in the Shops.

This artificial operation therefore affords a beautifull and christalline vitriol of VENUS, and some small quantity of Gold and Silver; whereof although there be not so great plenty that it may bring profit, yet it shews the possibility of the thing, and teacheth that a transmutation of Metals, may be exercised in any small Cottage by any Country Man.

But if any one shall collect Urine, and extract

from thence the volatile spirit of its Salt by Distillation, he shall far more deeply pierce the heart of the Copper, and shall obtain more of Gold and Silver than he can get by the but now mentioned rustical labour, of which matter more things are found in other places of my Writings.

LI. After what sort pure Gold may be extracted out of any Copper.

In all Copper a spiritual occult Gold lies hid, the which in the labour of separation by Lead in a Cupel or Test, is not taken notice of, or considered. But this very Gold is nothing else but the colour of Copper, so firmly adhering unto its body, that it remaineth very difficult to be separated from thence, but the operation being rightly instituted, although it be not gainfull, yet it demonstrates a possibilty of the thing, for the sake of experiencing whereof, to wit, whether there be Gold in all Copper, thou must labour after the following manner.

Dissolve Copper in AQUA REGIS, and pour much water on the solution, that by this water the dissolved Copper may be largely diffused into this solution diffused by the water, pour Lead that is dissolved in AQUA FORTIS, so that one or two half ounces of Lead may answer to one pound of Copper.

But as often as the Lead shall settle to the bottom, thou shalt shake or stir it, that the solution may be well mixed, and that the Lead may attract something of Gold; and may precipitate it with it self

to the bottom, the which being dried and separated in a
Cupel, will leave a small quantity of Gold, not an
argument of profit, but a token of a possibility, which
testifies that there is Gold in all Copper. But if any
thing of profit were to be received, such Copper was to
be taken which was already changed into vitriol. But
among vitriols the HUNGARIAN, CYPRIAN, INDIAN, JAPANICK,
and other the like do excell, which do offer themselves
in the Mines of Gold-bearing Copper, and are handled
after the following manner.

LII. The manner of extracting Gold out of natural
vitriol.

     Dissolve thou vitriol in common water, and pour on
the vitriol dissolved, a little of dissolved Lead, and
stir both solutions by frequent motions, that the Lead
may attract the Gold out of the vitriol, to be dried and
separated by a Cupel, which will remain like a grain of
greater or lesser quantity, as the vitriol shall contain
more or less of the same.

     Of the vitriolated water out of which the Gold was
extracted, thou shalt again make vitriol by so long
boiling, and evaporating till a thin skin appear, which
being then exposed to the cold will shoot into Crystals.
But this operation brings with it more profit, if it be
not now boiled up into Crystals, but that vitriol only
be taken which is drawn out of its Mineral into Water,
and after the extraction of the Gold, is at length
boiled up into vitriol.

     But least this fishing of Gold out of vitriol

should seem wonderfull unto any, we will shew a way, whereby Gold may be extracted out of Sea Water, or Sea Salt, the solution of Lead assisting: The manner is as followeth.

LIII. After what manner Gold is to be extracted out of Sea Salt, or Sea Water, not indeed with profit, but only that it may be demonstrated, that Gold is hidden even in Sea Water or Sea Salt.

Fill a great Copper Kettle with Sea Water, and pour therein a little dissolved Lead; the which goes to the bottom because it cannot indure Salt, and is straightway precipitated into a white powder, move and stir the water in the Kettle often, that the solution of Lead may every where touch the Sea-Salt-Water.

Through this action, a spiritual Gold adheres to the leaden powder, and sinks to the bottom together with it, which powder being freed from its Salt by common Water, and dried, and melted in a Cupel, leaves a small grain of Gold, as a remainder.

N. B. For this fishing Silver is more fit than Lead.

LIV. How, out of poor Mines of Copper, from which no profit can be perceived, Copper, as also Gold it self if it be present, is to be easily and without costs, extracted and separated.

The sandy or sulphureous Mine or Mineral of Copper is to be roasted or calcined, by burning even to the consuming of the Sulphur, because sharp waters do not assault sulphureous matters, the Mineral being calcined and beaten into a powder, full a gourd, and pour our solving secret on the same, the which I have taught above at the extracting of Minerals.

The whole dissolvent in abstracting or distilling is recovered, and that indeed not without increase. But the Copper, and Silver do stick fast in the dissolved Salt, which remained in the Mineral after the abstraction, and the which is to be washed out with water, out of which water, which drew out that Salt, the Gold contained therein, may by the solution of Lead or Silver, be drawn forth.

But if the same water be boiled until a thin skin appear at the top, and exposed to the cold, it will shoot into a green vitriol, but for the extracting the Copper out of the Salt-Water, Rods of Iron are to be put therein, which do attract the Copper, the which being withdrawn and washed clean, and melted into Copper by fusion, is administred for other uses: For because it is like to a tender and filed powder, it is changed by an easie business into Verdigrease, after the manner which shall by and by be taught.

LV. After what manner Gold may be by an easie business by Fire and Salt, be separated out of Copper.

If Gold shall be mixed with a mass or lump of Copper, all the Copper is to be reduced in a bright burning Fire into Ashes; and the Ashes are to be infused in our secret ACEUM or Vinegar, the which dissolveth the Copper only by decoction, and leaveth the Gold undissolved, like a shining powder, to be dried and melted with Borace, out of which Gold of twenty three Carats proceedeth.

That Vinegar, our secret ACEUM, draws all the Copper from the golden CALX. The Copper is separated from the Vinegar by Rods of Iron, being laid therein as we have taught in the foregoing manner.

LVI. How Copper being extracted out of vitriolated Water, and adhering to Rods of Iron, is to be changed into Verdigrease.

This pure Copper may be moistened with the strongest Vinegar, and put into earthen Pots, the which being covered with its Cover well fenced with Clay, are to be placed in Horse dung, and to be left therein for a time, yet so as that the heap of Dung be sometimes renewed. All the Copper is in a short time changed into Verdigrease, and indeed far more pure than that which is set forth to Salt in the Shops, and which is made in SPAIN, by the husks or pressed out of clusters of Grapes.

N. B. In extracting Copper out of Mines, regard is to be had unto this thing, to wit, that with poor and wild or course veins of Copper, LAPIS CALAMINARIS, or ZINK is sometimes found to be admixed; which is no ways perceived to be in them. But if those Minerals are extracted with AQUA REGIS, and this be to be taken away by Distillation, none of the AQUA REGIS goes forth, but only a flegm without savour, because the LAPIS CALAMINARIS or ZINK doth retain all the Acrimony with it, just as if those two Minerals should say to the AQUA REGIS, we do not as yet let thee go, because as yet, we have need of thy indeavour for our amendment; & etc.

But it is certain that whatsoever Minerals and Metals do retain with them sharp spirits, are as yet immature, and may be ripened by those spirits, that they may bestow Gold and Silver, as hath been already said, and shall as yet more largely be spoken to.

LVII. Out of wild or course Minerals, or veins of Lead, admitting of no melting, out of which no good Lead, much less Gold or Silver, can be drawn, how to extract not only Lead, but also Gold and Silver with profit.

As we have said above, that some Minerals or Veins of Copper do appear in Mines, the which by reason of LAPIS CALAMINARIS or ZINK, do refuse all melting, and can be by no fire reduced: So also we here admonish, that Minerals of Lead are found, the which do indeed contain much Lead, but by reason of the LAPIS CALAMINARIS, ZINK, and a sulphureous Sand being admixed with them, they cannot be overcome by any melting, for these matters do take away a ready flowing from the Lead, and do cause that such Minerals, which for the most part together with Lead, do also hide not a little of Gold and Silver, are cast away as altogether unfit, and unprofitable, whenas notwithstanding very much profit might be received from them after this manner.

Let the Mineral by pounding be broken in small pieces, and in my little secret Furnace which I have fitted for the calciñing of Minerals, let it be

roastedwith bright burning Coals, that the gross Sulphur
may conceive a flame, and burn. If in time of operation
the matter should gather it self into heaps or knobs,
and in co-heaping should make round Pellets, it being
taken out of the Furnace, let it again be beaten, be set
upon live Coals and roasted, and these labours be so
often repeated, until all the Sulphur shall be consumed,
and the Mineral doth no longer co-heap it self into
knobs, but being made bright burning hot like dead
ashes; it no longer sends forth a sulphureous stink. At
length out of these ashes being well washed, a dead and
unprofitable matter separates its self from the good and
metallick earth, the which being melted by it self in a
Furnace called by the GERMANS STICHOFEN, becomes a
flowable Lead which containeth Gold and Silver.

But if the Mineral be so stubborn that it
altogether refusing all melting, could not by it self be
reduced, and nevertheless contain Gold and Silver,
something of Litharge is to be added to that metallick
earth, which procures a flux unto it, and yields that
Gold and Silver bearing Lead, which by the common
operation wholly refuseth to offer it self.

LVIII. Another way teaching by the help of Salt and Fire
to draw Silver and Gold with great profit, out of all
stubborn or rude and untamed metallick earths, in whose
Veins Lead, Copper, Gravel, or course Sands, Iron, or
LAPIS CALAMINARIS have for the most part conjoined in
Society, and which do deny all profit by vulgar operations.

As Fire burns up every gross and combustible Sulphur in Mines or Minerals, that these do at length subject themselves unto melting, and do render Metals easie to be hammered; so also Salt fixeth, and makes constant whatsoever volatile body endeavours to flie away into the air, that it may afford a ripened, melted, and profitable Metal. For that cause such Minerals common Salt being added as was above-said, are to be roasted in live Coals, that that devouring gross Sulphur may vanish by burning with a flame, and that together also the Metal it self may be promoted to maturity, and so that by this very thing, good Gold and Silver may be separated, whenas notwithstanding otherwise, not any one should obtain so much as the least thereof out of these very Minerals.

Such an amendment and changing the more imperfect Metals into the more perfect ones, may be attained by the help of Salt and Fire.

If therefore common Salt, and gross Fire are able to perform this in Minerals, what shall not these not common but secret Fires of Salts effect, in trans-changing Metals already pure, into more pure and subtile ones?

LIX. After what manner Metals are to be amended by pure Fire, or the fiery spirits of Salts.

It may easily be perceived if a gross Salt and Fire do some good to more gross Metals, that also a more pure Fire and Salt may do more good on purer Metals. Instruments whose edges are made very sharp by whetting, are far more fit for operation than those that are dull, and will perform more. By how much the sharper an Auger or Wimble is, by so much the sooner it boareth through the Wood, and on the contrary, by how much the more blunt it is, by so much the slower it pierceth through.

He that is earnestly desirous to obtain any good in the amendment of Metals, he must of necessity apply the subtile and strong spirits of Salts, that he may destroy Metals with the same, may kill them, and reduce them into their former life, and so may procure more noble bodies unto them. When their former body is restored to those moist and cold Fires of Salts, to wit, that they may return unto the form of Salt, but of a more noble and subtile one, Metals may far more speedily be destroyed; a double Fire performs more than a single one, since therefore Salt is by it self no other thing but a meer and concentrated Fire, and the Fire of Wood or Coals joining it self with the other a greater efficacy must needs be expected from them than by common Fire alone, but we have hitherto made mention of such operations, and therefore its needless here to repeat them. From what hath been hitherto said, every Chymist may gather and learn those things which concern the amendment of Metals, wanting the help of Salt and Fire; more God willing shall follow.

LX. Let us now ascend higher, and demonstrate what incredible miracles or wonders our secret Fires of Salts may effect nigh to that great work of Philosophers.

As in the foregoing Chapters it hath been sufficiently confirmed, that unripe Metals may by the help of Fire and Salt, be particularly promoted to a more perfect maturity: So also in the multiplying of Animals and Vegetables, that thing evidently appeareth; to wit, if sufficient meat and drink be administered to any Infant, that he groweth dayly in bigness, and strength of body, until he come to the age and perfection of a Man. The same multiplying in Vegetables offers it self to our view, in that a small seed or root do snatch to them their nourishment from an earthly Salt; and the beams of the Sun, and do rise up into a perfect, great, and fruitfull Tree. This particular transmutation is conversant before our eyes, and therefore is a thing most known, but after what sort the most noble part or purest essence is to be extracted out of the bodies of Vegetables, Minerals, and Animals, that other more weak bodies may be strengthened and amended by the same Philosophers have always hidden and covered with the greatest endeavour. Hence it is, that there hath been very few, and as yet are, who have had the knowledge of this highest Science.

As to what therefore belongs to the great work of the Philosophers, all the Philosophers do in their writings with one accord affirm, the which I do also in all my writings affirm to be most true, and do as it were show with my Fingers, to wit, that nothing in the nature of things doth effect a Tincture and Tinge with a most gratefull colour, but Sulphur alone, and that one only, and that this same combustile immature and

volatile Sulphur is fixed and changed by the operation
and help of Salts into a true Tincture, the which is as
certain as that which is most certain, and yet labourous
also, and requiring a space of time long enough,
especially if any one doth insist in a moist way. The
way of coming unto the end of such a work as I think,
yet with the safety of others judgement, this is
the best, if any one bind or fix such a Sulphur, which
was already brought unto a perfect maturity by nature,
they might bring this profit with it, that it should not
require a longer time for its maturity.

But such a fixed, and tinged Sulphur, is no where
more nearly found than in Metals, and especially in
Copper and Iron, but the better and more pure in Gold;
the finding out whereof notwithstanding (by reason of
its most firm and intimate conjunction with its body, as
also its separation) hath been always esteemed almost
impossible. For unto diligent searchers, a true
separator which might separate the pure part from the
impure, hath for the most part every where been wanting.

For as it is known such a hard or compacted
metallick body, can very difficultly be separated and
dMded into its parts.

The solution of sharp waters, sups up indeed every
Metal, but it effects no separation. For because Metals
are Homogeneal things, and the metallick Sulphur is so
strictly bound to its metallick Mercury, by the bond of
the metallick Salt, it can never be brought to pass,
that by such solutions, or by precipitatings, or by
other ways, one part should be separated from the other.

If a Metal being dissolved by a water be precipitated all its parts being so mixed as they were before its dissolution do fall down and settle, and admit of no separation. But if any one would also render Metals spiritual, that so the more pure parts might be disjoined from the more impure ones by distillation, yet there is no separation made, but the more pure body it self ascends, and again as before, it consisteth of three principles, performing indeed more in Alchemy and in Medicine than the more gross bodies of Metals, but is unfit for a true Tincture, because nothing operates in all bodies, but a lively Soul, and that which vMfies or quickens other dead bodies, for it is the spirit, as Christ saith, which quickeneth, the body is unprofitable. Let man, or any other living creature be for an example, the which as long as it lives, it moves it self and operates as long as the spirit, the Authour of Life is present with it; but that vanishing away the body wants all motion, and remains a dead Carcass. If now it could be brought to pass, that we could lay hold of such Animal Spirits, and could render them corporal ones, we might also perform incredible things by the same, and perhaps fashion or form a living creature of a lump of earth, the which notwithstanding God hath reserved to his own self. But this that bountifull Father hath granted unto us, that out of unmoveable subjects, or those wanting a moveable and animal life, we may extract their pure Souls or Essences, and render them corporal, and effect thereby things of great moment in Medicine and Alchemy.

But the souls of Metals do excell herein, as being more fixed and constant than the essences of vegetables, but they are far more difficultly attained. For the souls of vegetables do suffer themselves to be easily extracted, but the colours and souls of Metals do hardly admit of extraction, and for that cause are accounted of by the ignorant for a thing impossible to be done, nor indeed is it altogether, without some cause; for the separation of the tinging soul from the hard metallick body is a thing of great moment; many are the ways that have been attempted for the procuring of this Sulphur; and some ignorant fellows have written Books, of the acquiring or getting of the same, whereas 'tis evident that they never saw such a Sulphur.

The most learned and most witty HELMONT wrote egregiously concerning this Sulphur; but yet not so clearly as that any one could out of those his writings get a perfect knowledge of the same. Nor indeed is it expedient that such kind of Pearls should be cast under the feet of swinish Men. There is no Writer (as far as I know) that hath mentioned any thing concerning this matter, clearer than ISAAC HOLLAND in his CH. DE AMANSIS, where he teacheth, that he who hath gotten the art of changing Metals into transparent Glasses with their peculiar colours, hath purchased a notable secret in Metallick affairs. He alludes (by way of likeness) to the bodies of Men brought to a clarity or brightness after this life, and thus declares his Doctrine and says; The souls of Metals do shine forth through their AMAUSA'S, or clarified bodies, clad in their proper colours, even as the Souls of Men shall hereafter shine in the other World, from (or through) their clarified bodies. And further he saith, that when such AMAUSA'S (or Glasses) are reduced into their former bodies; then the AMAUSA'S OF COPPER & IRON become fixt, that of Silver becomes Gold, and that of Gold becomes Tincture. The said Authour hath not clearly expressed the manner of accomplishing this, but in my opinion (without prescribing ought to any one) this is the nearest way of attaining to such an operation, viz, of getting the souls, or the pure Sulphurs of Metals, viz. If the metals be first reduced into AMAUSA'S or transparent Glasses, out of which their souls are easier extractable than out of their gross bodies. But now, for such an extraction here is such a MENSTRUUM required as doth not work upon all the whole body, or dissolve it, but doth only attract there out of the colour and purest Sulphur, and leaves the body behind white. But where shall we find the description of such a MENSTRUUM? None speak of it openly, but many mention it obscurely; nor indeed is it so necessary that such an ARCANUM of so great moment be manifested to every one.

But however, this in brief you are to be admonished of, that like draws its like and extracts it. If a rnercuriality be to be extracted out of the metallick masses (or bodies) then 'tis expedient to use a mercurial MENSTRUUM, for like rejoiceth in its like. So sulphureous essences are extractable by sulphurous MENSTRUUMS, and not by mercurial ones. For Water doth willingly associate it self with Water, and Oil with

Oils. And forasmuch as all the Philosophers write that the Sulphur or tinging Soul in Iron and Copper, doth as to goodness and nobleness equalize the Sulphur in Gold, it will be needless (in my opinion) to take Gold, but to bring Iron and Copper to that pass, that they may become transparent Glasses, from which their colours may be extracted. But if so be that any one is minded to prefer Gold before these, and to extract the tincture hence from, he may do as he please, and will find in many places of my writings a manuduction, (directing him) to the transmutation of Gold, (and so of the other Metals too) into transparent bodies, which thing is highly necessary. For there is not an easier way of extracting the tincture out of Metals, than by first reducing them into transparent AMAUSA'S. Now the MENSTRUUM serving for this extracting of the Sulphurs out of the metallick bodies is to be so prepared, that it dissolve not the body, but extract only the Sulphur or pure Soul thereout of. Such MENSTRUUMS PARACELSUS himself makes mention of, and affirms, that with them the skie coloured Saphyrs, the red Rubies, and the yellow Jacyrith may be so deprived of their colours, as that there remains no more Of them save only the white bodies. Besides, that white Crystals may (by the help of tinging sulphureous spirits) be died with various colours. Verily 'tis a secret of most mighty concernment, to have the skill how to prepare such a MENSTRUUM as will penetrate the most hard Stones and most compact Glasses to extract them, and withall to communicate to othersome various colours, without the corrupting, breaking and destruction of any of the bodies, the which thing seems in my opinion very likely to be true, though as yet I know not how to do it. That which I have tryed by my operations I can write and teach, viz, how all the Metals may very easily (yet one more easily than another) be changed into fair transparent and most delicately coloured glasses, and how out of these glasses the pure and tinging metallick souls may be extracted, viz. by such MENSTRUUMS as are sulphureous, subtile, not dissolving, but only extracting.

But forasmuch as these kinds of MENSTRUUMS are not (as far as I know) described by any one, and yet are the producers of such notable effects; I could not omit the discovering of something concerning them, for the sake of such as are Students in true Philosophy, and after some sort shew that kind of extraction, which is to be accomplished by the help of our concentrated spirits of salt, or of our moist Philosophical fires.

LXI. How a vegetable subtile sulphur is to be so actuated by the nitrous moist fire, that it may extract the fixt sulphur of metals, or their pure tinging soul.

First of all, the oils of the vegetables are to be exalted by distillations, and often repeated rectifications to the highest degree of purity and subtilty, and afterwards to be once rectified by some concentrated nitrous fire, that so being already of themselves subtile, they may get a fiery vigour endued with a faculty of seizing upon metallick sulphurs, and of extracting them out of the hard and compact bodies. For any vegetable oil how subtilly soever it be prepared, hath not any power of entring into the metals, and much less of having any ingress into their glasses, nor can it extract in the least, though such bodies should be covered with it for a long season. But now if an artificial operation shall have sharpened such an oil with those most subtile spirits of salts, and have rendred, it more acute and penetrative than those concentred spirits of the salt do lead in the sulphur, and bestow on it a power of attracting to it self its like.

And albeit that such concentrated spirits do when PER SE, and alone, wholly dissolve metallick glasses and make no separation at all, no, nor do not extract the least particle of any sulphureous substance, yet the case with them is castly altered, when such fiery spirits are artificially united with most pure sulphureous oils; insomuch that they bestow on them a faculty of working upon metals, and of extracting from them a most pure sulphur; nay farther, they purchase these oils a capacity not only of extracting the colours out of metals, but also out of other things. Now we have taught at large in our second part of Furnaces, the manner of rendring the oils of vegetables subtile; and as for the nitrous fire requisite to this operation, the way of concentrating it is to be found in this Book a little before, so that 'tis wholly needless to repeat the same things over again which have been afore treated of. And thus far is my knowledge come, viz, how by the help of oils animated and actuated by the aforesaid means; to extract the most pure soul out of metallick AMAUSA'S, or those hard and glassy subjects: But how such a sulphur is to be brought into a tinging medicine, I (professing not my self to so great a master) do not as yet know. Neither is it at present needfull to exceed the due bounds by so large a treating of such worthy things; for things wonderfull may be effected by this MENSTRUUM both in Medicine and Alchemy, and in other arts, concerning which we shall (God willing) presently add somewhat more.

And whereas I have made mention here of clarified bodies, and concentrated spirits; I judge it worth while also to shew what difference there is betwixt those concentrated spirits and clarified bodies as the Chymists call them.

The clarified bodies therefore are nothing else save bodies purged and mundified by the operation of the fire. For the fire is the ultimate examiner of all things, as being endued with such a power by which it burns up all things, reduceth them into ashes and powder, and out of the ashes makes glass; that being the utmost or ultimate thing whereunto all things are reduced. If therefore there be in any thing any good, which being burnt in the fire is by fusion or melting, turned into transparent glass, it doth manifest it self in its utmost or ultimate essence, and shine forth in its brightness, insomuch that every one may see what lay hidden in its life afore thus burning it. For example, I take wood, an herb, or an animal, I burn it and transmute it into ashes; these ashes I melt and turn into glass. Having thus done, there appears no colour visible, for the glass is white, and that because the sulphur is consumed in the burning; and the mercury is fled away into the air in smoke, as being two principles which are no ways able to resist the force of fire:

But the salt, as being a contemner of the fire remained in the earth of no efficacy.

But now a metal being by the fire turned into ashes, though part of the sulphur and mercury hath in the combustion and vitrification flown away into the air, yet notwithstanding the best part remains; and this is the reason why such metallick glasses are coloured according to the metals nature and property, and which (afore thus burning it) was hidden. We will yet farther evidence it by an example.

If I add to burn iron or CROCUS MARTIS the glass of lead, then the glass made by melting will have the yellow colour of a Hyacinth. The same CROCUS MARTIS being molten with common glass, made of wood-ashes and salt; yields a greenish coloured glass which is the natural and proper colour of the iron. For the lead altered the colour of the former mentioned glass of the iron and made it yellow in the melting, and so hindred it from manifesting its true and natural colour. The glasses of two several colours being molten together do exhibit false colours, as may be seen by co-melting a skie colour and a yellow glass, the which being molten together yield a green colour, and doth so represent it self both in the fire and out of it too. From hence took I occasion to write and teach the way of finding out (by molten glasses) what kind of metal is hidden in any mineral or metallick earth. Which way of proving mines or minerals is far better and speedier than that which is usually done by a decoction and exhalation of lead in the Cupel. Thus may you mix five, six, eight or ten grains of some finely powdered mineral, with one or two lots of Venice glass being of easie fusion, and put the matter thus mixed in a well covered crucible, and by melting it reduce it into glass. The colour which will be in the said glass, will shew what metal the MINERA contained: Lead will yield a duskish colour, tin, a white, copper, a Sea—green; iron a somewhat greenish, silver a yellow, and gold a skie coloured; each of which colours is the true and internal colour of the respective metal. Gold doth also resemble a Ruby as to colour if other colours be added there-unto. But yet in the mean time, the skie colour is its proper and natural colour, and so is yellow of silver; and this is notably agreeable with the truth, though to such as are ignorant, it seems a thing wonderfull, for indeed such mens knowledge ends in external things, but they are wholly ignorant of internal ones. But now the colours of gold and silver are better and more perfectly known, if there be added unto them some fix and white sulphur, which prevents the gold and silver from being thoroughly reduced into their peculiar bodies by fusion. If the CALX of gold or silver be molten with Borax, they both return into their former bodies, and do not pass into glass any ways coloured: But that some glass of easie fusion be mixed with these CALX'S, together with a little powder of flints and so molten, then the flints will (by reason of their sulphur) hold with themselves the gold and silver and so keep them that they admit not of fusion (or reduction) in their whole body, but do remain in the glass with some part of the metalline property which renders their internal colours visible, which else would not appear to sight.

N. B. If you have the mineras of gold and silver at hand and melt them with glass, their colours will also appear, because that in the minera's there always is some sulphur that hinders the metal from wholly returning into a body, so that some part of it abides in the glass, and therein shews its colour. This also is to be minded, that if haply some minera or metallick earth contains not one metal barely, but 2 or 3 more metals, then always that metal of which the most quantity is in the said minera doth after fusing shine in the glass beyond the rest. As for example.

Suppose I would make tryal in the red Granates (stones) I powder some eight or ten gr. and mix them with one lot of white Venice glass finely powdered, and I melt them, and so turn them unto glass. Now in this transmutation the glass doth not become red, but of a delicate grass colour, and so teacheth me what metals are hidden in those Grariates, viz, copper and iron, and also more of this, (viz, the iron) than of the other. And though there should be some gold too, yet is it unperceiveable because of the predominancy of the iron over the copper and over the gold: For in this operation that metal only manifests it self to sight, which is in greater plenty therein than the rest be.

ISAAC HOLLAND would by this vitrification signifie unto us, that after this life, viz, when the would is consumed with fire, there shall arise from the bodies of men reduced into ashes other clarified bodies, and of such and such colours, according as their souls have (either good or bad) framed, or as it were made unto themselves in this life-time in their gross bodies. What other thing (I pray) are fair colours, but the virtues of those subjects out of which they emit or send forth their splendour.

Take a similitude hereof from the melting of minerals, wherein though a mineral of silver or copper hath in it much silver or gold, yet if the superfluous sulphur be not (before the melting the said mineral) separated by a little as 'twere roasting fire; but be (together with that gross sulphur) set in a vehement melting fire, there will not verily be any metal gotten hencefrom, but that stinking sulphur would transmute the good metal into black Scoria's. So likewise, no fair and transparent glass can be by melting) made out of pure metals, if that kind of gross sulphur should adhere unto them.

These few things touching clarified bodies, I could not pass over in silence, and much less could I omit this, viz, that the bodies of all things may be much better transmuted into clarified bodies by our secret fire, than by the common fire. For the common fire drives away the volatile parts, whereas on the contrary, our fire doth preserve them and renders them

107

fixt and transparent as well as the other parts. And therefore of necessity these bodies must needs shine with fairer and brighter colours than those others, in which the common fire hath expelled the mercury and sulphur, and left remaining nothing else but the salt.

But now as concerning such a transmutation into ashes by our moist and secret fires, any one may easily guess the way. For whatsoever is put in them must be necessarily burned into ashes, and they far better ashes too than are made by burning in the common fire after it the common fire burns any herb or wood into ashes, the sulphur burns away in a flame, the mercury betakes it self to its wings and away it flies, and the salt abides behind in a few ashes, or a little earth. Now our Philosophical calcination takes away nothing but conserves all (the principles so called) together; and doth in the first place produce to view a black coal, then afterwards other various curious colours, and then a white colour; and at last to complete the operation, it yields a red fusile and medicinal stone.

N. B. Here it is to be noted that for preparing a pure medicine, a pure subject is to be made use of; for if to be that any one would endeavour the transmutation of an herb, wood, or any animal into a medicament by the help of the secret fires, then all the ashes and feces which were in the herb would also adhere unto the medicament and would render it impure, therefore necessity requires that you do not take the whole herb, of the whole animal, but only their essential salt, the which being void of feces consists only of the pure

principles of the herb; and doth easily admit of being
transmuted into a red tinging, and more soluble stone
than the herb it self with its feces by it, doth.

I would not have you to account of these things
here delivered you as if they were of small moment. No,
for they are such things as cover over with this their
vile or base covering, such matters as are of great
weight, and which will not come to every bodies
knowledge. Surely 'tis a considerable thing that a part
of any vegetable, animal or mineral body should (by
conserving all the most volatile parts, and by rendring
them altogether constant and stable, without the least
loss of weight) be ripened into a fixt soluble and
tinging red, and medicinal stone. This way of transmuting all things without loss of the weight thereof
into clarified bodies, is of all others the best. And
those bodies on this wise clarified are without doubt of
greater efficacy than are the gross bodies themselves of
the animals, vegetables, and minerals, which do as yet
abound with their gross and impure feces.

But if so be any one be not herewithall content
but panteth after higher things, be may advise with
himself about extracting the soul out of this red and
fixed stone, and reduce it again by a reiterated
operation unto the form of a stone, whereby he will
without doubt make it yet far more effectual. And by how
much the oftner any one shall repeat this same
operation, so much the more effectual a medicine will he
obtain, for it will at every reiteration notable augment
its virtues, for by such actions the efficacy and

virtues of things are con-centered and driven into a
very little compass, wherewithall wonderfull things may
be performed.

We are yet moreover to see what spirits are, but
principally what concentrated spirits are, and what they
are able to do.

And because the matter in hand gives occasion of
treating thereof in this place, we will briefly give you
a declaration of the same.

LXII. What spirits are, and by what means they operate
good or evil.

In the first place, there are spirits called
vegetable ones, viz. When vegetables are beaten to
pieces, and being confused are moistened with water,
(provided they have not juiciness enough of their own,
or do altogether want it) and so termented, being
fermented they are to be distilled, which distillation
brings forth subtle and efficacious spirits, and such as
are the effecters of many profitable things in Alchemy
and in other arts, besides the use thereof in medicine.

Secondly, sundry and divers spirits are also made
out of animals by distillation, as out of Blood, Urine,
Hairs, Horns, Hoofs, and such other parts of animals,
also which spirits have their use in Medicine and
Alchemy.

Thirdly, there are also spirits which are
expelled, or forced out of minerals and metals by the
force of fire, but principally out of Salts, as Vitriol,
Allum, Salt Peter, common Salt and such like; of the

preparations of which kind of spirits, the Books of
Chymists are full, and therefore stop us from the
superfluous repetitions of the same. But as for the
concentrating of them and the utility of them, it hath
been already described by us.

I hereby give occasion for all the diligent
searchers after true Medicine and Alchemy; to
contemplate, what may be effected in Medicine and
Alchemy, if those fugitive spirits were, by our fixing
and moist fires which separate not any one part from
each other, but do fix all the parts together, deprived
of their volatility and made fixt. These few things we
were willing to mention concerning the spirits which are
subjected to a Man's power and are within his reach, and
serviceable for the use of mortal man.

LXIII. The particular medicinal use of the con—centrated
spirits of Salts.

We have heard that the concentrated spirits or
moist fires of salts do reduce all things into a CALX,
after a Philosophical manner, without a forcing away of
the mercurial part, and a burning up of the sulphureous;
insomuch that (by conserving, altering and bettering)
they fix the whole. Being therefore compelled by a love
to my neighbour, I have a mind to set down in this place
some medicaments, as well universal as particular; but
yet so as that they may not fall into the clutches of
(my) unworthy enemies, but may be reserved only for
friends.

And first, here shall be a medicine mentioned,

that amendeth the weak digestion of the Stomach.

Take out the teeth of a Wolf or a Dog when he is half dead, (being first shot with a Gun) and pour thereupon two or three parts of the concentrated spirit of salt in a Cucurbit, set the glass upon sand that it may be heated moderately, whereby the oil of salt may dissolve them and bring them into a thick pulse or mash. Upon this mash pour warmed water that so all the Acrimony of the spirit of salt may be separated therefrom, and that there remain only the white pulse, wherewith (because some of the oil of the salt doth yet remain, and is not washable off wholly with water) a Sugar candy is to be mixed, that so that remaining Acrimony may be allayed, and the pulse be the pleasanter for your use.

LXIV. An Antidote against Poison.

The teeth of a mad Dog being prepared after the aforesaid manner, do yield an Antidote against Poison. And indeed so do the teeth of all Animals, but especially the teeth of Stags and their Horns, do (after such a preparation) resist all Poisons.

LXV. What Beasts they are whose Teeth and Horns do (as a medicine) exceed the rest.

The teeth of all ravenous animals, as likewise of Sea monsters, and their Horns, are of great use in medicine; as of WOLVES, BEARS, LYNXES, TIGERS, LIONS, LEPARDS; and as to the watery animals, CROCODILES, and such like ravening fishes, whose horn, teeth, and scales, and likewise the claws of Birds of Prey, may be by the concentrated spirit of salt, converted (after the aforesaid manner) into good medicaments.

LXVI. An experimental discovery of what Vermine are fit for use of medicines.

Take strong and well rectified spirit of salt, or only a strong AQUA FORTIS, put thereunto your Vermine, of what kind soever it be, and it will presently endeavour to get out as soon as ever it feels the said moist fire. But being it cannot get out, it will struggle till it dies. Now by how much the longer the worm or flie liveth in the AQUA FORTIS, by so much are its virtues in medicine greater, and this may serve as advice to every one. As for the way of preparing medicaments out of Bark or Trees and Husks shall be mentioned by and by.

LXVII. An experimental discovery, of what Herbs are profitable for Medicine, or unprofitable.

You are to make use of the way but now mentioned concerning Vermine, and such Herbs as thou knowest not, or such whose virtues are to thee unknown, put (one after another) into the moist fire; and that Herb which is of a slower solution, excells that which is sooner dissolved, both in strength and virtues. For example, Lettice, Purstarie, Mellons, Cucumbers, and such like waterish Herbs, (and so is it with fruits too,) have a moist nature, and are presently turned into water by those fires. But Rosemary, Sage, Thyme, Dodder, and other hotter Herbs, require a longer time for their solution. Ginger, Pepper, Cloves, Nutmegs, Cinamond, Cardamoms, Zedoary, and etc. do require a yet longer time as to their solution, afore they will thoroughly pass into a water. From hence may any one know the nature and properties of Herbs very easily. This also is to be observed, that the Medicine out of a Vermine, or out of any Herb, is by so much the more efficacious, by how much the vehementer venenosity it abounds withall.

LXVIII. The mariner of preparing an effectual medicament out of venemous Vermine and Insects.

I have taught in the second part of my spagyrical PHARMACOPAEA, a way of correcting venemous insects by the fixt Liquor of Niter, and of transmuting their veriome into an effectual Medicine, which (way of preparation) he who is studious of good medicaments will there find. But now in this place is taught, by what means such like Verinine, and such Herbs as abound with Venome are to be corrected by the concentrated fire of salt, and to be turned into excellent and penetrative medicaments. The operation whereof is thus.

Pour into some glass vessel, one, two or three ounces of our concentrated fire of salt, then put thereunto such Vermine as you would prepare your Medicine of, one after another, provided that you do not put more in, than the said fire is able to dissolve and consume. When all are dissolved and converted into water, all the poysonousness is lost (or gone) and they become good medicaments.

LXIX. The manner of separating the medicament made of
Vermine dissolved by the moist fires.

There is found to be a great difference amongst
Vermine and venemous insects. For some of them are of a
dry nature and property, some of a moist, some of a fat
and oily nature, insomuch that it is altogether needfull
to make a due distinction of their natures. Such insects
as be of a dry nature as Canthandes and such like, are
to be used in the form of a salt. The aqueous Vermine,
as Earth-worms, Spiders, and such like; they exhibit
their medicinality in the form of a Liquor: The
Balsamick Vermine, as the May-worms, and others of that
kind, do (beside the medicament they afford) yield also
a fat and Balsamick Oil; and indeed (both for external
and internal medicinal use) much more effectual than the
Liquor it self.

But that the thing may be the better understood, I
will here set down an operation, which every one may
follow as a leading Star.

LXX. How the operation in dealing with all kinds of
Vermine is to be used.

Take some ounces of May-worms, put them in a
glass, pour upon them so much of the concentrated spirit
of salt, that the Worms may be well covered therewith
and be by little and little dissolved; after that they
are wholly dissolved, put the solution into a separating
glass, shut the mouth of the glass with your finger,
then turn the glass upside down, keeping it so long shut

with your finger, till all the fat oil swim at top of the Liquor. Then take away your finger that the Liquor may run out, and when the oil comes, shut the mouth of the glass again with your finger, and let it run out into another glass. Keep this Oil or Balsam as a precious treasure, with the which thou wilt perform wonderfull effects in the curing of diseases, but principally in the Gout and Stone. But yet thou wilt get but little oil from these Worms, and when you put them into your dissolving Liquor, you must have a care that you do not touch them with your hands, but take or catch them with a small Forceps, and so put them into your glass. For they have in them this property; if you touch them with your hands, they presently colour them with their fat Balsom that they cast out, which somewhat resembles the smell of Musk. As if they should say, PRAY LET US LIVE, FOR WE GIVE THEE ALL WE HAVE: TAKE THIS BALSOM AND MAKE USE THEREOF FOR THE CURING OF INCURABLE DISEASES.

Some men studiously carefull in such affairs have gathered this Balsom, and have found it to be far more efficacious than the Worms themselves, yea indeed too strong, because they were ignorant of the way of correcting it.

LXXXI. The separation of the medicinal Liquor from the moist fire, after the separation of the Oil.

As concerning the Liquor from which the Balsom is separated, viz. the medicinal parts is very hardly separated without a mortifying of the moist fire;

therefore the moist fire of the Salt is to be killed
with a contrary fire, that so the separation that is
required may be made; and 'tis thus to be effected.

Filter the Liquor consisting partly of the
dissolved Worms, and partly of the fire of salt, that so
it may be rendred clear, and free from the Coals or
Husks of the Worms, if haply there be any of them
remaining undissolved. And if (by reason of the too much
fatness it be very difficult to filter, pour thereunto
so much common square Crystals. The same doth oil of
vitriol and salt of tartar. But the salt that is in this
operation made of the common salt exceeds the others, as
to sweetness. That which comes from vitriol, doth beget
a nauseating by reason of its bitterness; and that which
ariseth of salt peter is of a middle nature. But yet
they do all three of them enjoy a laxative and purging
faculty; and likewise provoke Urine either stronger or
weaker according as the Vermine are, which these said
fires have been used in the dissolution of.

N.B. That the operation of the Liquor doth always
exceed the virtues of the salts. When the sharp spirits
of salt are not mortified with a LIXMIJM of Tartar, but
with the spirits of Urine or of SAL ARMONIACK, the salt
and liquor become far stronger, than when the
precipitation is done with Salt of Tartar. For the
spirit of SAL ARMONIACK doth for the most part
precipitate the dissolved and corrected Vermine into a
Powder, which being washed off with common water, and
freed from all Acrimony or sharpness is used in
medicine, in a dry form. But this, the LIXIVIUM will not

do, but always conserves the Verniine in the form of a liquor.

LXXIII. Question. Whether or no there may be any more or any other usefull things learned from this solution of venemous Vermine?

For Answer. Yes, for this operation doth not only teach the good and bad properties of all Vermine, but doth withall evidently demonstrate, that every animal of what kind soever, (yea and Men themselves too) when they are put into such an agony and perceive the approach of death, do discover and clearly evidence the internal motions of (their) nature, which they have (in their life time) been indued withall.

LXXIV. The way how to know the internal nature of every Worm in the earth, Fish in the water, Birds in the air, yea evon of Man himself.

Take a SCARABAEUS or Beetle, either such a one as lives in Horse-dung, or else one that is of a coppery colour, put him into a glass wherein is some AQUA FORTIS, and you shall see that in the utmost necessity (or last agony) of death, he will not endeavour to get from out of the AQUA FORTIS, but will strive to hide himself in the earth according to his innate property. But whereas the bottom of the glass is too hard for him to get through, he will be so long endeavouring to accomplish his desire, and in strMng to get through the bottom till he dies. From hence may it be perceived what his ultimate refuge (or shift) is; viz, to endeavour the shurining of his approaching death, by sheltering himself within the earth.

If you put a flie in the AQUA FORTIS, she will not go to the bottom but will do her utmost to get out at top, because her living is in the air, and so all volatile or flying things are wont to do. As for a fish if it be put to its shifts, it endeavours to shun the danger by betaking it self to the bottom.

In such a kind of manner doth the nature of men become apparent, when they are reduced to the extreainest of difficulties. A Godly man, whose thoughts are in this life time always upon God, will constantly adhere unto him in his Agony, and being upon dying will betake himself to him for hiâ refuge, in whom he hath at all times built his hope, and waiting or looking for help from thence, from whence he hath always hoped for it.

But the ungodly Man who hath never in his life time feared God, nor set him before his eyes, but hath always yielded himself with his thoughts unto the will of Satan, he will very hardly Implore (in his greatest anguish) the help of any other than of him, to whom he hath (in this life time) adhered in all his thoughts and actions.

LXXV. The preparation of good medicanients out of venemous vegetables, by the concentrated spirits of salts.

We have hitherto taught, that our moist fires of salts do indeed destroy all things, but do not burn up, or force away ought of such things as the common fires are wont to do. That this is true, even the vegetables themselves bear witness, which being put into our moist fires are therein dissolved, and pass into a water. But their oil which is in them is separated, and swimnieth at the top, and so may be separated thencefrom, as we have mentioned above, concerning the May-worms. After the same manner the essence of the herb may be severed from the spirits of the salts, as we have there declared. The oils which by this operation are drawn out of the herbs and other vegetables, do obtain great and peculiar virtues, because they are excellently well corrected by the efficacy of the moist fires, and are amended, which correction, and bettering they do not at all attain by their being distilled and expressed.

LXXVI. The correction of the too vehemently purging subjects by the moist fires, whereby they may be safely made use of.

DIAGRIDIUM or SCAMMONY, HELLOBORE, CATAPUTIA, GAMBOGIA and other vehemently purging subjects may be dispoiled of their venemous faculty by the aforementioned way, and be rendered more sweet and milder.

LXXVII. The correction of the too vehemently operative Diureticks, whereby they may be of safe use in the cure of the Stone.

Dissolve Canthandes, May-worms, Earth-worms, Millipides or Piglice, in our concentrated fires, and follow those ways of operation which we have afore prescribed, and you shall acquire an excellent and safe indicament, having a faculty of healing the Stone of the Bladder and Reins.

LXXVII. The amending of narcotick and somniferous subjects, by our moist fires, that so they may perform or shew their virtues without hurt or danger.

Take OPIUM, Heribane seed, Mandrake, or the like subject that provokes to sleep, pour on it the concentrated spirit of salt, and it will melt (or dissolve) therein; if there be in it any oily-ness, as in the Henbane seed is usual, it will separate it self, and swim on the top of the liquor, the which is to be severed from the liquor, and to be warily kept; because it being only anointed on the Temples will presently cause sleep. The liquor is to be used internally, as we have prescribed in the precedent preparations.

LXXIX. The amending of venemous subjects, that are together purgative, suderifick, diuretick, and somniferous, by our moist fires; insomuch that they do not only become safe, but are the effecters of much good in medicine.

Amongst the number of such kind of subjects, are Stavesacre, or the seed of the louse killing herb, Levant Berries, vomiting Nuts, and many others of such a like faculty, which are to be proceeded withall after the same manner, and by the same operations as the former.

By this or the like way may all venemous, and vehemently operative vegetables and animals be corrected, so as to be safely admitted to internal uses, and to be producers of such effects as are of great moment in Physick; whereas otherwise (though they have in them excellent virtues) they cannot by reason of their vehement operations be taken into the body without danger.

LXXX. Whether or no poisonous minerals may be corrected as well as the vegetables and animals, by our secret and moist fire of salt, and be turned into wholesome medicaments.

You are to know, that not only venemous animals and vegetables but likewise all the minerals that abound with poison may be amended, and their most present or speedily operative poison be converted into most excellent medicines. For example.

LXXXI. How the venenate and volatile minerals are so to be inverted by our moist fires, that the volatile be rendred fixt, and the poison be made a medicine.

Take of white or red Arsenick one part, pour thereto two or three parts of the concentrated fire of niter, the which (niter-spirit) you shall distill thencefrom in a head and body in sand; then take the remaining matter out of the glass, and wash it with common water, which being done, you shall have the Arsnick, Diaphoretick, and such as may with safety be taken into the body; but yet in a small dose, because it doth sometimes provoke vomit, and principally when the nitrous fire is something of the weakest. But to prevent such vomiting, the said fire is to be twice or thrice drawn off from the Arsnick, by an Alembick; that so the poison may the better be slain, and the volatility transmuted into a fixity; and the same Arsnick may be afterwards molten and handled with the other metals without all danger of poison, which was impossible to be done afore. For the Arsnick whose poisonousness is not as yet removed from it, cannot be admitted into the body without great danger. Neither do we here insert the preparation of such medicaments out of Arsenick, and the like veneinous minerals, for this cause that they should be introduced into medicine; no, for there are other safer medicaments to be had, our aim herein is only this, to shew that even the most poisonous, and most fugacious or volatile minerals may be inverted or turned in and out by our moist fire, and dispoiled of their venenosity and rendred fixt.

LXXXII. The manner of transmuting the fugacious and easily fusible (fluxible) minerals by the moist fires of salts, so that being fixed they hardly admit of fusion or melting.

   To exemplifie this, let us consider of Tin or Zink, which are reckoned amongst the metals of most easie melting, and are most volatile. For the vulgar know that Tin is molten with a very little fire, and doth thereby vanish in fume, if it be but kept in continual flux. But if it be calcined by continually stirring it into ashes, it becometh fix, nor doth it admit of reduction to its former body by any violence of fire, but is turned into glass.

So after the same manner do we roast or calcine Tin, ZINK, and the other flying metals with our moist fires, and burn them into ashes, and they such ones too as do not return to their former body, and thus 'tis done; when we pour on them our fiery liquors, so as that they heat together, or do even by abstraction (or distillation) again free the said metals from those liquors; for then these metals remain in the bottom like dead ashes, nor do they suffer themselves to be by any means reduced to their former bodies.

N. B. But whoever he be that knows the using of such matters and powders thereunto, as can reduce such ashes to their former and fusile bodies, such a one will not spend his labour in vain; for he will get a metal of a much more noble and better nature than Tin, whose greatest internal part is gold and silver.

But yet let no body imagine that he can perform this reduction by the help of Borax or Salts; no, in no-wise. For there are metallick matters required to this labour, to cause a fluxing or melting, sundry preparations whereof I have taught to and again in my writings, but not under such a title or name as if this power or efficacy of thus doing were ascribable unto them. For I have barely mentioned their use in other metallick transmutations.

LXXXIII. How flying mercury is to be so fixed as to

admit of heating red hot.

Coagulate common mercury with common sulphur into a black ashes, and then mix this ashes with the concentrated fire of vitriol, or rather with such a fire as is extracted from sulphur it self; so as that there may arise from this mixture a thin mass; of which mass put one or two lots in linnen or cotten rags, and so rowl them up that one fold may come over another, and the mercury may be in the middle. Then tie this ball firmly and strongly with a thread, and let it be environed all about with the fire, that so all those rags may be red hot and changed into Coals.

Take out all this red hot mass, let it cool, separate the burnt linnen rags, and you shall find the mercury turned into a red powder; but yet it hath no ingress into the metals, nor performeth it any thing of much moment in medicine, because it is converted by the burning of our fire into a dead earth. Neither have I here mentioned this coagulation as if any gain were to be received thence-from; but only on this account, that the most great virtues and powers, of our fires may by the operation thereof be demonstrated.

LXXXIV. Another experiment easily demonstrating the possibility of rendering mercury constant in the fire, by our secret fires of salts, which thing the known and common fire can never do.

Melt one part of common and yellow sulphur in a crucible or earthen Pot, and being molten like oil cast thereinto two parts of common mercury, and mix the

matter well with a SPATULA, that the sulphur may assume
the mercury, and may pass with it into a black mass. To
which mass you must yet add so much sulphur as the
weight of the whole mass in the Pot is of. Then all is
to be molten together, and to be by well stirring so
long mixt until it get an ashy colour. Then you are to
dip in the said mass as it is in flux, pieces of linnen,
such as they are wont to use about fuming their Barrels
with a brimstony odour, to preserve them from stinking.
Such rags being put on an Iron Crook may be kindled, by
which kindling they are burnt up, the sulphur and part
of the mercury vanishing into the air, but some part
being calcined with the flame of the sulphur, and fixt,
sticks to the burnt linnen. But what virtue this
ealcined mercury abounds withall I cannot tell, as
having never experimented it, and I have only inserted
here this operation, for this end, that the power of our
moist fires may be made apparent. Many more meditations
and inquisitions will be thereby laid open, which otherwise would never have been sought after nor found out.
For in this labour there operateth a twofold fire, viz,
the visible flame, and invisible moist fire which the
sulphur hideth, and by the burning up of its body,
manifesteth, and renders visible and efficacious.

For that heavy acid oil of sulphur, which sticks
hidden in all sulphur, performeth most great matters in
metalline things; because the flame or external fire
exasperates and forceth on the internal, sharp and moist
fire of the sulphur, that acid quality sets upon that
subject that is adjoined to it, and destroyeth the same,

and advanceth unto a more fixed state.

But that I may in some sort satisfie the greedy searcher after truth in this thing, it seems to me expedient here to mention the occasion, which (without studying thereafter, viz, this coagulation of mercury) did by chance bring me thereunto; and did chiefly shew me an excellent ARCANUM of bringing all Wines, Vinegars, and other such like drinks easily and speedily to a clarity and transparency. Such a secret it is, as I believe never was known as yet to any, and therefore worthy to be here set down for the common benefit of mankind.

LXXXV. An historical discovery of the reduction and restoration of tenactous and corrupt Wine, to its former clarity and goodness.

I had some Wine in a Vessel that became viscid or ropy and tenacious; for the amending of which, I sent for a Wine cooper, he pours it out as is the usual custom into another fresh vessel, and used thereunto all his art, that so he might better it. He passed it oft times through a long Pipe made of white plates, and perforated with many small holes, and many other means he used, even whatever he had knowledge of, but yet all he attempted was in vain. Then at last he put into the Wine no small portion of salt, and shook them both together very strongly, but all in vain, insomuch that he left my Wine corrupt (as he found it) and out of all hope of restoring it.

But because there was too much salt thrown into

the Wine, yea so much as that it might be perceived even by the taste, the Wine was rendred unfit to be drunk, though it should have (thereby) recovered its former clearness. So there remained nothing else to be done therewithall, save the extracting of its spirit by distillation. Yet nevertheless I had a good mind yet to try whether or no it could possibly be freed from that tenacity; to this end therefore I kindled some sulphurized rags, being sprinkled over with the mineral or ore of lead reduced into powder, and with that fume imbued I my Wine, as is the usual custom when Wines are through corruption degenerated from their good state or condition. I added thereto the ore of lead because that as the sulphur was burning the fume of the lead might penetrate the Wine and precipitate all the defilements to the bottom. But this experiment did not fadge. Then I took mercury and mixt it with sulphur after the aforeshown manner (in the foregoing Chapter) and dipt some rags therein and kindled them, hoping that the mercury being transmuted into fume, would have ingress into the Wine. But yet it appeared quite contrary in the use, the sulphur indeed was consumed by burning, but the mercury would emit no fume, but was contrarily turned into a red powder, and stuck on to the burnt rags. After these burnings, (viz, of this murcurialized sulphur) often reiterated, the Wine did not smell of the Brimstone, as its usual to do, but of Musk or Ambergrease, and recovered its former clarity; but yet not fit to be drunk because of the overmuch quantity of the salt thrown thereinto.

Thus it happened unto me, the which thing others may consider of with a more accurate meditation, and may haply (from this history) apprehend such things, as may in other matters be very profitable. For it is not in vain that I mention these things in this place. Enough is said to the wise. 'Tis sufficient that I have showed the way, if any one refuse to go in the same let him blame himself.

LXXXVI. How our moist fires of Salts are able after a sort to fix the yellow and common sulphur, so that it may be used with profit both in Medicine and Alchymy.

Take one part of yellow sulphur beaten into powder, and four or five times so much in weight of the concentrated fire of salt peter, which spirit pour upon the said powder in a glass cucurbit, and extract or abstract it thencefrom several times by an Alembick; this done, the sulphur in the cucurbit will get a red colour and become pellucid or transparent.

If it resolves in the air into a fat oil, the operation is well handled; if not the labour is to be repeated either with the self same fire or with more new, which is the better way. For the said fires are to be so often drawn off thencefrom until it flow (or resolve) into a fat oil: An oil I say of sulphur which is endued with great virtues, not only in Alchymy and Medicine, but may likewise be used in other arts with a great deal of profit. But especially it is an egregious Balsom, resisting all the sicknesses of the Lungs, and other putrifying corruptions, as shall be apparently evidenced in the following Centuries, more largely treating about these things.

LXXXVII. A way of turning Antimony into a snow-white medicament, by our moist fires of salts, and which is of safe and profitable use against the Plague, all Fevers, and other diseases.

When the REGULUS of Antimony made per se without iron, and beaten into a powder is persued or throughly moistened with the concentrated fire of salt peter, and is for a while kept in warm sand; the moist fire burns the REGULUS of the Antimony into a white powder. After that the whole shall be of a white colour, pour therein common water, and it will imbibe or draw to it self the fire of the niter, which will again be fitting for other labours, and perform the office of spirit of niter.

The white powder being by many washings rendred sweet, and then dried, performs the office of sri excellent diaphoretick medicament and may with safety be used; it strongly resists the Plague, all fevers and other diseases, that are to be expelled by sweat.

LXXXVIII. By what means black and crude Antimony is to be reduced by the nitrous fire into a white powder, and the combustible and yellow sulphur separated therefrom, that it may serve as a PANACAEA for the resisting of all diseases, and may operate by the four Emunctories, Vomit, Stool, Sweat, and Urine.

Antimony is by so much the better and nobler, by how much the longer and fairer Rays or Stria's it appears to be of, and therefore such is of greater efficacy in medicine than all other sorts. To this therefore being powdered, pour so much of the nitrous fire as may serve to dissolve it; the which fire will presently even in the cold, begin the work of its dissolving. When that is done, put the glass in warm sand that all the Antimony may be dissolved, and its yellow sulphur may swim at top of the solution like a yellow powder. The solution being cold, strain it through a pure linnen cloth, and the sulphur will stay behind in the cloth, and hath its peculiar use in Medicine and Alchemy. But to the solution pour common water, thereby to quench and weaken the nitrous fire, so that the Antimony may fall down to the bottom, in the form of a tender and snow-white powder; the which being well washed and dried, may be made use of as a PANACAEA to drive away many diseases: For it operateth with a singular efficacy by all the Emunctories, but yet very safely, unless any one doth too foolishly and unskillfully abuse the administration: It is also endued with all those virtues that I have ascribed to my red PANACAEA.

LXXXIX. By what means the concentrated fire of Kitchin salt drives over Antimony in a retort like Butter, and affordeth a matter of profitable use in Medicines and Alchemy.

Pour upon the REGULUS of Antimony beaten into a powder, the heavy concentrated oil of common salt, the which being again drawn off thencefrom in a retort by distillation, brings over with it as much of the REGULUS of the Antimony as it can, and ascendeth like a thick Butter. It is a mighty fire, and very fit for the ripening of some immature metals; and withall is most profitable in Surgery, and lays a good Basis and foundation for the curing of incurable and cancerous Ulcers. If you pour water upon this oil, the Antimony precipitateth out of it, in the form of a white powder; and is to be afterwards well washed and dried; so that being reduced into that white powder, it becomes a good medicament to be used in all those diseases, wherein the aforesaid medicaments are appliable; but with this caution, that it be given but in a very small dose, because it is of a more powerfull operation than the aforegoing medicamentS prepared by the help of the nitrous fire are of, and that for this reason, because the oil of common salt makes things fugacious or flying, but the nitrous fires renders them more constant in the fire.

XC. The way of turning mercury into red, and strongly purging medicament by the operation of the nitrous fire.

Abstract two or three parts of our nitrous fire, from one part of purged mercury, by distillation in a glass cucurbit; and it will make the mercury far more fixt than if AQUA FORTIS were many times drawn off therefrom by distillation. This red mercury is to be freed from its saltiness with common water, and so becomes a strong purge, and is to be used in a small dose of one, two, three, or at the most four grains, and effecteth the cure of MORBUS GALLICUS, and other such like loathsome diseases.

After the same manner there may be easily prepared, not only sundry and excellent medicaments, by the operation of our moist fires, but there may be likewise done things of great moment in Alchymy and other arts; concerning which time will not permit me at this present to make a more ample narration of, but I will remit it to the next following Centuries.

Now forasmuch as I call in this Treatise the concentrated spirits of salt, moist fires, and yet as to their outward shape they represent no shew of fire at all; I deem it necessary to shew by a more firm demonstration, that they abound not only with fiery virtues, but also are (after their inside is turned outwards, and their outside inwards) true, visible, palpable, and sensible fires, but especially the nitrous fire, which best of all confirms this our opinion and saying, it being prepared by the Chymick Art and operation out of a fiery subject.

XCI. The way of converting or turning the internal and yellow colour of our moist and white nitrous fire from the inmost parts, outward, and making it visible.

That there is hidden a yellowness and redness in niter, is not beyond the reach of any ones capacity, but it is very easily likely, and credible. For seeing that salt peter is a solar child, it must necessarily answer to (or resemble) its father the Sun in colour, form, virtue and efficacy, if it would purchase belief with any one as to its original and natMty. But salt peter shines with a white colour, but the Sun is clad with a yellow garment and shines like the fire; insomuch that there is no correspondency or likeness of colours, though otherwise there is found the greatest similitude in burning, and in ripening all things. This only being the difference between salt peter and the Sun, the one, viz, the peter doth particularly only augment, ripen and advance all things; but the Sun doth universally bestow on all things, life, increase or growth, and nutriment, but yet the salt his companion is an helping assistant as shall be evidently proved in the end of this book.

I do verily believe, that if it were an easily accomplishable thing by us, viz, to extravert the internal and innate redness in salt peter outwards, and to separate the same from its unclean and gross body, and knew we how to render it fix and constant, we should perform things of most great moment, in an universal way.

But yet for the removing of this doubt, I will shew that salt peter is the son of SOL, though (as to its outward Physiognomy, it resembles not its Father. I do therefore say, that its Father is the yellow Sun, from him it is generated, but yet by the help of the white Moon, she is the Mother, and bestows on it the white colour. But I say, that the paternal blood and fiery virtues it hides in its inmost bowels. So wisely is Salt Peter signed by its Parents, viz. by the Sun its Father, and the Moon its Mother. The Father bestows on this, his Son a fiery heart; the Mother a white and cold body; from hence 'tis that it is clad with an Hermaphroditical nature, being Male and Female together, hot and cold, red and white, vivifying and killing.

XCII. Of the admirable nature of Magnetism, attracting to it self its like.

According to my simple and small judgement, the red colour of salt peter is not (by the operation of any other thing to be separated from its white body, better and more commodiously than by the affinity and likeness of some certain magnet that will touch it.

For example: Let us consider a little of common gold and Common quicksilver, they are so linked with a tye of mutual love, and internal likeness to each other; that the one draws the other unto it self.

For if in dealing with mercury any portion thereof should happen to fall on the ground, and dashing it self into a thousand Atoms, it cannot be by any kind of way better gotten together again, than by the help of such a magnet, as will attract to it self the dispersed and dilated Atoms; such a magnet metals are, and especially gold, as being conjoined to it in the nearest affinity; therefore I sweep together this so widely dispersed mercury, together with the earth and other defilements from which the said mercury is scarcely distinguishable as being all over covered therewithall; and to these defilements do I put a piece of copper, silver or gold, which being well shaken to and again amongst these filthes, draws to it self the widely dispersed and small Atoms of the mercury, and so recovers it by extracting it out of all that rubbish.

Now when the metal hath attracted mercury enough, and can attract no more, the mercury is to be wiped off from the metal with a linnen cloath; which being again well shook amongst the trash as you did afore, draws to it self the other Atoms; these labours are to be so often repeated, till it be all extracted, and so renders it thee the same without any loss.

Just so and after the self same manner may the inmost soul, and which is largely dispersed throughout the whole body of the salt peter, be extracted thencefrom; were only such a magnet but known unto us, as had a great affinity with the soul of the niter.

I will yet set down another, and a more evident similitude, that so the business may become the more clear and manifest and be the better understood.

XCIII. A clear and evident demonstration, whereby is shown that even the most hidden things may be manifested and rendred visible by their magnets.

Let the admirable nature and property of the common magnet be well considered; nature having endowed it with two plainly contrary virtues, one of attraction, the other of expulsion. For on one of its sides it draws iron to it self, and on the other of its sides it drives it off; and thus it does, not only in its great pieces, but also when 'tis broken into very small bits. For always on one side it draws to it self the iron, and on the other sides drives it from it self, by this operation respecting both poles, viz, the Northern and the Southern.

But to return to my purpose; I will demonstrate by evident examples and operation, that the inmost and most hidden nature and properties of things, are wont to be most evidently manifested and obtained, by attracting, and repelling magnets. For all the things that are, have their enemies and their friends, as shall be proved in the following operations.

XCIV. An operation demonstrating or affirming, that the internal and hidden natures and properties of things may be manifested and obtained by attractive or repulsive magnets.

Dissolve some lots of lead, and such as is wholly void of silver, in AQUA FORTIS, and precipitate the lead by pouring thereunto some salt water, this (precipitated lead) wash with fair water and dry it. Take some three or four ounces of this CALX of lead, and thereto admix a fifth part of pure gold, being first reduced into most pure and most subtile Atoms, on such wise as hath been taught in other places of my writings; but if you have not at hand such a CALX, use another CALX of gold prepared any kind of way, but yet the first CALX is the fittest for this operation. Melt both the CALX'S, viz. the leaden and golden one in a crucible, that the lead may become a fusile stone; but the gold CALX will (by this operation) be much heavier, and be white, this whiteness is nothing else save pure and good silver, drawn out of SATURN by SOL sympathetically, and made visible, which (afore) lay hidden in the lead in a

137

spiritual and invisible manner.

But some may here demand; forasmuch as there is so much silver hidden in all lead, whence comes it that none are found that get it out from thence? I answer, that there are indeed a many that would get out great masses of silver out of lead, did they but understand the art, and could so bring it to effect. But seeing they are ignorant of the natures of metals and their properties, and know not how to do any thing, they cannot become masters of their wishes. Now in this extraction, there is a two-fold cause presents it self, viz. Sympathy, and Antipathy. The gold by reason of the kin and amity it hath with the lead, draws thencefrom unto it self the spiritual silver; and because of an inbread hatred it has to salt, it drives away the same from it self. The gold therefore (in extracting the spiritual silver out of the lead, hath an assistant, aiding it by a contrary operation, and so bringing to pass, that there is so much the more silver extracted, because the salt being added to the lead, doth by reason of that inbred enmity and difference 'twixt it and silver drive this, (viz. the silver) away from it as its enemy.

And although that out of such lead prepared with salt may silver be always molten, yea without adding any gold thereunto, meerly because of the inimicitjousness that is between silver and salt, whereby it caused that the silver is thrust out of the lead by the same as by its enemy; yet so much silver is not gotten by that way, no not by the half, as is drawn out by the addition of gold.

For when the gold attracts, and the salt expels, there are made two actions together, the one by Sympathy, the other by Antipathy, both aiming at this mark, viz, to extract the hidden silver and gold out of the lead.

Let there be evaporated in a cupel two small centenaries or hundreds weight, each of like weight; and to one of the centenaries add some pounds (of the proportionable small weight as the centenaries are) of pure gold, and there will come from that centenary more silver by the help of the gold, than from the other, whereto was added no gold: But the gain by this operation will not be much, or haply none at all; but this is only to shew, that it is verily possible, for gold being put upon the cupel with lead, to get some silver thereout of, and to be more in weight; which effect is produced only by a sympathetical faculty. Be now if salt doth also lend to gold its assistance, then is there a twofold operation of a double operator; whilst in extracting of the silver out of the lead, the gold is occupied in attracting, and the salt in expulsion.

These things were of necessity to be laid open by me, forasmuch as they teach by what means the inmost and most pure parts are to be separated from the more gross; so that every one may know the natures and properties of things themselves, viz, with what love they imbrace each other, or what hatred there is betwixt them.

By this experiment then, may any wise and

understanding man easily learn and believe, that even out of salt peter or (any) other white body, the red soul may be extracted. Whosoever therefore shall know how to get these helpers, viz. Sympathy and Antipathy for his purpose, shall never labour in vain, but shall at all times reap fruit by his labours.

And as we have shewn that out of any lead, by the help of gold, a good part of silver may be extracted; so likewise may the same be clearly proved to be done with the other metals, and which may also be effected without the help of gold. Yet nevertheless the more fixt metal doth more readily and willingly attract the more volatile and purer part of the other impurer metals, than an unclean metal doth, and even much more readily yet, when there is afore adjoined to that metal from which any thing is to be extracted, an enemy of that thing which you labour to extract.

Upon this account therefore was I willing to insert an experiment, that so none might account of the thing mentioned as of small moment, but rather that he accurately ponder thereupon in his mind, that so he may thereby arrive to things of great moment by a well examining of the same.

Now as it hath been clearly and evidently taught that fixt silver may be gotten out of any lead, as well by Sympathy as by Antipathy, even so may it easily be proved, that the spiritual gold may be extracted out of other metals, partly by Antipathy, and partly by Sympathy, but much easier by Antipathy and Sympathy jointly together, so that one matter may draw unto its self the object it loves, and the other may drive from it what it hates; as we have proved in lead. If then this may be done in metals, why must it not be likewise done in other subjects.

We will therefore proceed on and see, whether or no it can be so brought to pass, that the hidden redness may be drawn out of salt peter by Sympathy and Antipathy.

Having therefore understood by the things already spoken, that like draws to it self its like, and is repelled from its unlike, there remains nothing else for us to do but to know what that like is, by which salt peter suffers it self to be extracted.

When we advisedly consider the rise or birth of salt peter, it is not to us unknown that it draws its originality from the excrements of animals, but especially from the dung of horned Beasts, as Cows and sheep. And forasmuch as Sheep and Cows do feed only on Herbs and Grass that grow in the Fields, and that those vegetables do proceed from the terrestrial salt by the help of the solar beams, it is more clearly evident than the Noon day light that the hot Sun is the Father of salt peter and the cold night the Mother, the earth the Nurse, and Salt the Food, nutriment and encrease of the same; the which is to be understood as in reference to the MACROCOSM or great World. But the vegetables, or all shrubs, herbs, and all grass which arise out of the earth in the MACROCOSM, cannot be more aptly compared with ought, than with the Hairs and Wool of Men and beasts, which are born out of the earth of the MICROCOSM or out of the animal body, like as the shrubs and herbs, and grass do spring forth and grow out of the MACROCOSMICAL Earth. Upon this score, the hairs, hoofs, or claws, and horns of Beasts; likewise the feathers and claws of Birds, and also the teeth and scales of Fishes, do altogether square as to their similitude with salt peter; they being such things as whereout of, together with other superfluous excrements of nature, true and good salt peter may be made with ease.

And like as to the procreation of vegetables in the MICROCOSM, and for their increase or growth, there is requisite a fat and salt earth, the warm splendour of the Sun, and the fruitfull Rain, whereby all kinds of fruits are born, and ripened; but contrarily by the penury or want of salt (it being the only nutriment) and of the warm Sun beams, and. of the Rain which is the promoter of all fertility, every thing that is vegetable must needs perish and die; even so is it in the body of Man: For as long as the heart of Man is in a prosperous healthfulness, and that the central fire, or vital spirit, and radical moisture be not defective, all things are well and in good equipage, and the whole body takes increase or growth, and the hairs grow plentifully: But on the contrary, when meat and drink fail, the whole body suffers loss, consumes and withers away and the hairs fall off.

But to comprise all in brief, I say, that all growths and augmentations as well in the MACROCOSM as in the MICROCOSM, must of necessity be at a stand and lessen as soon as the warm solar beams, together with the nutriment it self ceaseth and is deficient. So then it is a truth, that in Man as being a MICROCOSM or little World, and in the other animals, the hairs may be compared with the Trees, Shrubs, Herbs, and Grass of the great World, because of the great likeness that is between them.

And therefore the hairs of animals and hoofs, claws, feathers and scales of them are very like to salt peter, insomuch that one part doth after a sympathetical manner extract from the other, the most great virtues and inmost soul, and so one doth manifest the other.

For example, when the skin, claws hairs, hoofs, or nails of a man or any other animal, as likewise the feathers of Birds are smeared with the spirit of niter, or anointed therewithall, they presently become as yellow as gold, and do as twere put on a golden hue. It may now be demanded, from whence ariseth that colour? Comes it from the hairs themselves, or from the niter spirit? If that golden colour did arise from the hairs themselves, then it would of necessity be, that it should also discover it self, when the hairs are moistened with other sharp and strong waters; but thus twill never do, but only when they are smeared with the spirit of niter, or else with AQUA FORTIS, which containeth the niter spirit. From these things therefore it is evident, that the superfluities of the MICROCOSM have a most notable affinity with the superfluities of the MACROCOSM, viz. herbs, and grass. Hence comes it to pass, that one part draws or sucks from the other part its best virtues and power, and renders them visible, which afore lay hidden invisibly and impalpably in their gross bodies.

XCV. The manner of extracting out of niter its gold-like soul.

If we would go the nearest way to work with niter to extract its soul, then the gross niter is first to be mundified by distillation, then afterwards out of this purged body is the most pure part to be extracted by a convenient magnet, and the gross foeces to be removed; and this most pure soul to be brought by concentration and fixation to the utmost degree of perfection and dignity.

And albeit I could here set down in more clear expressions, the manner of extracting it, yet I am not so minded to do because of the unworthy. Let this manuduction suffice, whereby it is shown what means it is to be done by, viz, by some magnet attracting to it self its like by a magnetical operation. I can at all times exhibit such a yellow gold like soul of niter, and use it in the sicknesses of my neighbour. But, enough as touching these matters, we shall be more large concerning them, in the following Centuries.

XCVI. How the moist and cold fire of nitre is to be so ordered as to yield its visible flame.

Put some ounces of our concentrated and moist fire of niter in a glass, and pour thereupon drop by drop a sharp LIXIVIUM made of Wood-ashes, or rather of fixed niter, and keep pouring on so long, till all the noise, fuming, and ebullition cease, and that the moist fire it self be wholly allayed and slain. This done, all the corroding faculty is taken away from that fire, which said fire doth by this operation return to its former nature, and is changed into such a salt peter as it was afore its being converted into a moist fire. Out of this salt peter, being now made purer and better by so many conversions and operations, may a new moist fire be extracted by distillation and concentration, which is far better and much stronger than the former.

And now if this second moist fire be again extinguished with a LIXIVIUM of fixt niter, and be again turned into salt peter, and this peter be by a new distillation and concentration turned into a moist fire, this said fire will be endued with a far greater virtues: For in every mortification and vivification it becomes one degree stronger, nobler, and more efficacious; and so is the salt peter it self too bythose conversions and reductions exalted several degrees, and is at length brought to that pass, that it can do more wonderfull things than the common is wont to do; for one pound of such a salt peter being exalted to the utmost degree of subtility is far more efficacious than many pounds of common salt peter, and stronger, and much excels it in virtue. But it is not expedient that every one should know, what may be effected with such a most subtile and most pure salt peter.

The ancient Philosophers hid the preparation and use of common salt peter; and why shotild not we also hide such a salt peter exalted to the utmost degree of subtilty, wherewith the common peter is not at all comparable, especially in all such labours whereunto the common sort is wont to be applied, this operates much readiler, and far better and more effectually.

XCVII. An operation shewing the manner how by the help of Salt peter promoted to the highest degree of subtilty, the superfluous combustible sulphur of the imperfect metals may be kindled and burnt up; even as common fire burns up wood, insomuch as nothing will be left remaining save a little fixt Salt and ashes; so likewise in the burning up of the impure metals by our most pure salt peter, there remains also nothing save that fixt gold and silver which lay spiritually hidden in the metal, and is (now) left by the combustible Scoria's.

Every one knows that out of the common Salt Peter and Brimstone, may Gunpowder be made; but yet short in goodness, of that which is made of purified salt peter. By how much the purer and subtiler the salt peter is, so much the better and stronger powder doth it make. The same may be understood as touching the other uses of salt peter.

Further, every one knows that the common salt peter reduceth the common metals into a Scoria by burning them, and washeth gold and silver, and leaves them pure, concerning which fiery washing I have hitherto mentioned several things. But that the common salt peter doth perform this washing better than the pure, and this pure better and more efficaciously than

146

the purest, is no such matter in the least, which thing experience will most manifestly open to him who will try the same. Verily a small fire will not do those things which a greater will do, nor will the greater effect such things as the greatest will, and this is so evident and manifest that there is none dares deny it.

Take one part of the REGULUS of Antimony and four parts of pure Tin; melt them in a crucible and pour them out, and let them cool; this mass makes all iron and steel fusile, therefore when you would melt iron or steel, fill a crucible with either of the metals, set it in a Wind Furnace, and leave it so long in the Coals till all the matter wax highly red hot. Then take off the cover and put into the crucible, half as much of the said mixt mass of REGULUS of Antimony and Tin, as the iron or steel put in the crucible weighed, then put on the cover, and cover it over with the Coals, and urge the fire as strongly as ever possibly you can, that so the mass you put in, may cause the iron or steel to melt. When tis all well molten, pour it presently forth, least the Tin flying away in fume, leave the molten iron, and so it returned to its former hardness and not suffer it self to be fused.

This matter consisting of the REGULUS of Antimony, Tin, and Iron, or Steel, is so hard, that you may strike fire thereout of with a flint.

Now then if you would experiment the abovesaid combustion or burning up, take a good strong crucible made of potters earth, and fill it with salt peter, set it on live Coals so that the salt peter may melt, then having cast your tin and iron in the form of small rods, heat one end of the rods so as not to melt, hold the other end in a pair of Tongs, and put it into the molten salt peter, that the iron together with the tin and REGULUS of Antimony may be burnt up as if it were wood, and vanish away with the flame into a fume. For almost all tin and iron are a meer sulphur, and being consumed by the flame, leave nothing else in the crucible save SCORIA'S, which being washed with water, and boiled on a cupel or test with lead and blown off, do leave behind, the true gold and silver hidden in both metals.

For when by the flame of so pure a salt peter, the impure sulphur of the iron and tin is burnt up, it must necessarily be that what good soever was in the metals do remain behind.

I do not therefore here set down this operation, as if I would thereby promise any one golden mountains. No such matter. For I only manifest these, and such like labours meerly for this end, that every one may know, that salt peter being brought to a requisite purity, is wont to burn up imperfect metals as one burns up wood; and it may be easily gathered thencefrom, that such a pure salt peter doth as to its virtues much exceed the common peter.

As for such like labours of burning up the imperfect metals by salt peter purified in a due manner, and of getting pure gold and silver with profit, they shall be taught in the following Century (God willing).

For even as this first Century doth for the most part treat of fire and salt; so the chiefes-t part of the following Century shall treat of the wonderfull and great efficacy of purified salt peter in destroying, and reducing metals, and that with great bettering of them, and with no small profit. And albeit I was desirous of inserting in this first Century, such like profitable betterings of the metals, yet it could not well be done; principally because that there are many other things concerning the profitable use of the concentrated spirits of salt, that I must necessarily pass over here, because the number of this Century is already up, and therefore must I refer them to the following Centuries.

And forasmuch as there is frequent mention made in this Century, of glasses and crucibles, which none can be without in the preparing and use of these moist fires, because of the many hazards and losses wherewithall common instruments are accompanied, for they often break, or else easily let out or spill the boiling matters; it is altogether requisite that I should here have manifested this excellent invention of mine, which preventeth all such discommodities; and which I hinted at in the second part of my MIRACULUM MUNDI.

But whereas I have bestowed both those inventions there on the poor, of meer gift, it would be an unjust thing to take away from them what is theirs; nay rather they should have by good right more bestowed on them. So then being not able to proceed any farther as to this case, this thing only remains, viz, an affirmation that neither Medicine nor Alchymy can want or be without such excellent Inventions. But yet if any one desires to have them, he may write to those two men, to whom I have given them, that they may trade for the poor; whatsoever any one that is desirous of knowing those secrets shall bargain with them for, he will not be repulsed but obtain his desire, and purchase from them the secret; the which process I will nevertheless describe, omitting the naming of the matters.

XCVIII. The way of putting glasses in distillation and digestion, and so conserving them, that the boiling matter be not spilt.

TAKE with this matter fence your glass, that the matters you put into them run not out, or be spilt, and you shall not lose them.

XCIX. The manner of preparing such crucibles as will hold metals in flux a long time, and which can neither be broken nor melted.

R. Mix these matters and moisten them with common water, that they may be converted into a mass, of which you shall (by a crucible mold) frame small and great crucibles, knocking them into your mold with an heavy mallet, according to the manner I described in the fifth part of my Furnaces. Then take them forth of the mold and dry them in the air, and when they are dry use them; for they need not any burning in the Potters Furnace. They will (being rightly handled) stand a long time in the coals, and will not chap, neither will they melt with the most vehement fire.

C. An infallible demonstration, that in salt and

fire all things lie hidden; or, that by the help of the Sun and Salt all things are generated, arise, grow, and encrease.

Forasmuch as I caused to be stamped at the beginning of this small work, a circle with a square in its inside, and with these annexed words, IN THE SUN, AND IN SALT ARE ALL THINGS; (the truth of which figure and inscription is sufficiently enough evidenced by the so many operations described in this Century;) it seemed worth while unto me, to make an assertion here in the end of this Treatise by a true and evident demonstration, that all things are procreated, conserved, and encreased by the Sun and Salt, as being the principal and most noble creatures of God; but with this provizo, that there be present the seeds of those things that require multiplication.

For though the Sun and Salt were yet far excellenter and nobler subjects than they now be, yet notwithstanding could they not produce or generate so much as the smallest herb or meanest worm (to say nothing of a Man) without seed. If we have but seed, then it is permitted us by God, to propagate the same by the Sun and Salt. The beginning of seeds, God hath reserved to himself alone, The seed therefore is for us sufficient, which if we have, we are able to propagate and encrease it even to infinity, by the efficacy of the Sun and Salt, (that universal nutriment of all things).

The universal medicament and nutriment that the Country men use about conserving their Vineyards, Grounds, and Pastures, is the Dung of Cattle and Sheep;

from the which Dung being laid to the roots of the Vines and Trees, and thrown into the Fields and Pastures, the Vegetables do attract their necessary nutriment, do grow and bring forth fruits needfull for the support and nourishment of Men and Beasts.

But forasmuch as this nutriment which all the Vegetables do extract out of the Dung of the Beasts is nothing else by a urinous Salt, and that we know how to prepare out of the common Salt, such a urinous Salt which may be used about dunging and fatning the Earth instead of Dung, therefore verily we may be without that said Dung, being but furnished with such a Salt; the which being Alkalizated by the fire doth extend it self much wider, or goes farther, and dures much longer in the Earth, nourishing and dunging it, than the Beasts Dung doth.

Besides it gives to all Fruits and Corn, a far sweeter savour, and smell, than Dung it self doth, the truth of which will easily appear to him that will make tryal of the same. Nay more, and what is of far greater moment; there may (by the help of such Salts be communicated to Fruits, especially to those that grow on Trees, and to Grapes, a most fragrant odour; if instead of Dung, such Salts be used to the Fields and Gardens, and some spices or other things of a fragrant smell be mixed with those Salts, and put to the roots of the Vegetables. An example will illustrate it more clearly.

Plow up some part of some barren Ground, such as is so for want of Dung; or, even some meer sandy ground, and throw upon it of the said salt as much as is sufficient, and by plowing mix it with the Earth it self, and sow therein any kind of grain you please, and it will spring up out of the salted Earth or Sand, and be encreased, and recompence the pains taken about it, with an hundred fold encrease of the seed, even just as if you had dunged it, nay better than with the common Dunging.

And now I pray, whence doth that faculty of growing, encreasing, and multiplying arise, save from Salt alone, which you sowed your barren Field withall? For the seed cannot purchase to it self nutriment, and aliment, growth and encrease from the barren Sand and Rain water only. And this you may try the truth of very easily, if you fill two wooden Chests or Boxes with common Sand, and put to the Sand in one of them one, two, or three pounds of that inverted Salt, (according to the bigness or smallness of your box) and in the other box let there be only Sand. If now you sow divers seeds in each of the Sands, and set them in the air, moistening them with Rain-water, those seeds will indeed grow out of both the Sand boxes and get increase; but yet that which grows where it has been sprinkled with Salt is far fuller and perfecter, insomuch that it will hold on its growth till it comes to its utmost ripeness, whenas the other seeds in the other Sand box will by little and little perish and die.

It evidently appears from hence that the faculty of growing and increasing, in all things proceedeth from Salt only. If so be therefore, that Pease, Beans, Oats, Barly, Rye, Wheat, Wine, and all kinds of Fruits, do grow, and are increased by the benefit of Salt, it necessarily follows, that Salt is the subject and universal nutriment.

But that it may be understood how the common Salt (which otherwise is wont to corrupt or extinguish all the faculty of growth in those things it is mixed withall is to be inverted or turned in and out, that its hurtfull corroding property being put off it may put on the nature of Alkalies or urinous Salts, I do here covertly set down the manner of the operation, so that it may remain hidden to mine enemies and be communicated only to my friends.

Take common Salt, & etc. (See the Preparation in Appen. 5 p. PROSPERITY OF GERMANY).

Let them be well heated red hot together at the fire, this done, the salt will be dispoiled of its Acrimony and get an urinous property, and being used in a due quantity instead of Dung, to barren Grounds, causeth that the seeds sown therein produce much fruit. But this kind of dunging requireth frequenter Rains than that which is done with Dung; and therefore it may be better and more commodious in moorish places than Beasts Dung, if it may be done, which conjoins the said salt, and produceth the most wished effects.

But especially the said salt is far better for Vines and Fruit Trees than Dung Is; for it gives to Grapes, Apples, Pears and such like Tree-Fruits, a far better savour than Dung is wont to afford:

Besides, this benefit accrues to Vines thereby, that being sprinkled with that Salt, they have oft times nutriment enough for twenty years, and do every year bring forth Fruit most plentifully; whereas the nutriment of dunging will scarce suffice for five or six years. Further, That Salt may be far easier, and with much less costs

carried and transported into the high Mountains than Dung can; for a far greater quantity of Dung is required than of the said Salt, which said Salt diffuseth it self far larger, or goes much farther than the Dung, and dures longer.

I did this very spring time last past, put such an inverted salt to some half dead Vines, which being planted in a lean sandy ground could scarce grow up a fingers heighth, and they presently began to flourish, and grew up so fast that the growth was day by day perceptible. When therefore on a certain day I was shewing some Friends of mine the melioration of metals, in my Laboratory, to be effected by the help of common salt, as they were wondring at the thing, I shewed them those Vines afore my Laboratory, that sprang up from the half dead stocks, and they measured some of the branches of the Vines, found that in two or three months they were grown some 6, 7, 10 and 11 foot long, and the stocks or stalks whence they grew were two or three times thicker than at first; which great change or transmutation in the Vines, proceeding from the salt, would be more admirable and profitable than that of metals, if we did not look upon that transmutation of gold with such covetous eyes. And forasmuch as this transmutation spoken of, was observed in the month of JULY, and that there are almost yet three months for Vines to encrease and grow, any one may easily conjecture what an encrease the wood (or branches) of the Vines would have in the space of a whole Summer.

But let this suffice concerning the inversion of the common salt, being a most profitable promoter of the growing faculty in all things; the which things I was necessitated to declare here in the end of this small work. That so I might demonstrate those things which I asserted at the beginning of the same, viz, that in salt lie all things hidden, and by the seeds of things and the help of the Sun are rendred visible, palpable and essential.

An Admonition to the friendly Reader, or a Proposition not proscribing or tying to any body, whereby is shown how much benefit the Country may in general obtain by my not chargeable extraction of Gold and Silver out of the fugacious or flying Minerals.

That my inventions may be in many places profitable for the whole Country, I judged it worth while, briefly to declare my meditations or conceptions thereabouts; and to make a declaration of the way or manner which I judge it may most exceeding profitably be done by.

First of all, I suppose it is sufficiently well known that Princes and Noblemen are occupied or troubled with otherguess burdens and business than to employ or busie themselves with the care and enquiry after the metafline mine-pits, that are here and there in the Countries subject unto them. And if they should commit the care of these things to their servants, 'tis likely that they would rather pass their time in feastings or merry meetings, than in a painfull search of Mines and metallick Veins in the woods and mountains. And as for the subjects themselves, seeing they are altogether rude and ignorant of such arts, by what means, and with what success they can undertake such kind of Labours, any one may easily guess. These are the Causes why things of such great moment are let slip without any profit at all, and are plainly neglected. But in my judgement, though it be but slender, any Prince that hath many Subjects under his obedience, may every year gather store of gold and silver , and that without costs, if he would but only cause a small Laboratory to be erected, wherein the poor Mines may be extracted with due waters; and leave given to such of his subjects as breathe after the knowledge of such Arts, to frequent such a Laboratory, and there to learn such Arts, with this PROVIZO, that every one should bring the gold, silver and copper boiled or gotten out of the poor minerals by the help of the said extraction, into the Money-shops or Coining-house, at such a rate as they are every where esteemed or valued at, and not transport it out of the Country. Now by this means, not only the chief Magistrate would without any cost and labour get no small profit by the money, but likewise every body would to their utmost, labour in the inquisition after such poor Mines, that he might get Gold and Silver, and other Metals, out of those poor metallick Veins, and get gain for him and his. But now if the Prince or Magistrate will not permit his Subjects the exercise of such a work, but keep all to himself, any one may easily conjecture, that not so much as a man will set about searching after such Mines, hut will rather hide them, especially if the Magistrate (as is wont to be sometimes done) would constrain his Subjects to such kind of Labours.

This (according to my simple opinion, no ways prescribing to any what to do) would be the readiest way, without hurting of any man (nay rather it would help and assist many a man) of furnishing our Country with Gold and Silver Coin, which Foreigners have made it bare of. But this will not be by any way, unless the Magistrates themselves do make a beginning, as to the institution of such a work, by this means inviting and stirring up their Subjects to undertake such like Labours, which will bring great Treasures even to the whole Country. By this means may rich Mints be set up in very many places, instead of those which at this time afford Money or Coin so sparingly, and no small portion of Copper too, mixed with it.

These few things was I willing to advise for the sake of the good of the whole Country; only laying down my simple opinion without prescribing a rule to any, hoping, that no body of what rank or condition soever they be, will take it otherwise than well.

Secondly, Every Prince and great man would mightily promote the common welfare of this poor, if he would but take the care of shewing the way of so inverting common Salt by one hours heating it red hot, and bringing it to that pass, that it may be made use of instead of Cattles dung, for the fatning and bettering of barren Vineyards, unfruitfull Gardens, and other Fields that are backward or slow in bearing Fruits: for verily, even from this very Art would redound much profit to some Countries. For many Vineyards here and there, and many Grounds do want due tillage because of the scarcity of Dung, whereas otherwise they would prove very profitable both to the Subjects and Magistrates, if they could be fatned and made fertile by this kind of way.

Besides too, all Wines would be had in much more plenty, and be of a far sweeter and pleasanter taste, by such a medium, than if the Vineyards and Fields were dunged with Beasts dung. But as touching this thing, see more in the continuation of my MIRACULUM MUNDI.

The end of the First Century.

THE COMPLETE WORKS

OF

RUDOLPH
GLAUBER

trans; Chris. Packe

RAMS
1983

THE CENTURYS

SECOND

THE SECOND CENTURY

OF GLAUBERS'S

Wealthy Store-house of Treasures.

Which doth Illustrate his hitherto published Writings, with a more evident Explanation, and doth more clearly demonstrate the Truth hidden in them.

The First ARCANUM or Secret of the Second Century, SHEWETH;

By what means such Metals as are imperfect, wild, and in a manner unapt for use or sale, may be ripened or bettered by Common Salt and Fire, so as to yield forth Gold and Silver with benefit and advantage.

I have in the foregoing first Century, as also in the Appendix to the fifth part of the prosperty of GERMANY, described the incineration or reduction into ashes, which is to be done with Coals in a peculiar Furnace, fit for torrefaction or calcining.

Though this be a laborious way and tedious, yet is it not without its benefit, provided that a great quantity (as I have already oft-times said) of the not vendible minerals and metals be thereto used. But that such an incineration may be done after a more easie and compendious manner, the following way may be made use of.

Build an Hearth of good and fitting earth upon a firm foundation, put thereupon a Furnace (or an Arch) of good stones, adjoin hereto an Oven, (or side Furnace) out of which the fire may play and emit its flame over all the said Hearth, and pass thereout of by a chimny made for that purpose. Upon this Hearth put those metalline earths being broken in a Mill, and commix them with the Salt, and Coals reduced into powder, and leave them for twenty or twenty four hours, that they may be all well fired and heated red hot: For by this means, the salt makes the fugitive metal in some sort constant and able to brook the fire; and the wild sulphureousness leaves the metalline mineral, and adjoins it self to the salt, and converteth it into a vitriol or SAL MIRABILIS. This twenty or twenty four hours heat, gets a constancy and fluxibility to those wild metallick veins, and doth withall by that labour so fit and prepare the salt, that it doth afterwards by an easie mutation pass into good salt peter.

After that the said minerals have gotten themselves a better state by the said Cementation, they are to be drawn out of the Fire or Hearth, with iron instruments fit for such a purpose, and new and fresh minerals are to be put in, and to be dealt withall after the same manner as we said but now.

The minerals that are taken forth are to be broken in a Mill, and then the salt to be washed off with common water, and to be afterwards used about making salt peter, the which we have taught in the Appendix.

The light Coals (or Scinders) and unprofitable

162

earth is to be separated by water, from the metalline part, and this metalline part, or heavier LIMUS, being reduced and molten in the Furnace called STICHOFEN, yields a beautifull or pure and gainfull metal.

There are sometimes found in many places of GERMANY, wild, fugacious, and unmeltable minerals of lead, which for that they contain in them sulphur, Antimony, or LAPIS CALAMINARIS, do not admit of reduction in the Furnace called STICHOFEN, but do either go off in fume, or turn into dross. But being first roasted after the aforesaid manner, and fitted for liquefaction, the lead, comprehending in it silver goo, may be thence gotten with profit, whereas otherwise they are wont to be dealt withall without any fruit, and are therefore given over. This incineration therefore is profitably used to such degenerate minerals.

Now if so be any would deal with vendible and good metals, and would have profit from them by incineration, he must proceed this following way.

II. The manner of reducing lead into ashes, and so handling it with the spirits of salt, that gold and silver may be thence gotten with profit.

I have at large taught in my first Century, that in the ripening of metals and other chymical operations, a greater fire is endued with stronger power than a lesser, which is easie to be understood by those that have any wit.

I just now taught the maturation and bettering of unprofitable and wild metallick earths with crude and gross salt.

But forasmuch as the gross salt and a weak fire cannot of necessity put forth so much strength as a stronger fire is wont to do, therefore for such as desire a stronger fire than the common salt, the purer part is to be (by the help of Art) drawn out of the crude salt and to be separated from its grossness and impurity, the which is easily brought to pass by distillation, And as for these fires of salts, and the procuring them in great plenty, my writings, but especially the precedent first Century, do clearly and evidently treat of them, and this second Century will yet treat of it more.

III. The operation of incinerating the lead, or reducing it into ashes.

Having built a Furnace such as is for Cementation, put therein a strong iron Pot, just after such a manner, as the sand cupels (or Pans) are wont to be made, let there be a Grate to make a fire on, let the Furnace be bigger or lesser according to the bigness or littleness of the Pot you would put in, or according to the quantity of lead ashes you would make. Put the fire under the Pot and heat it red hot, and put thereinto so much lead as is requisite for the covering of the bottom of the Pot; the lead being molten, stir it about in the Pot without ceasing, with an iron spoon having a long handle, the which labour will turn the lead into ashes in the space of about two hours. Take these ashes out, and put in more lead into the Pot, and repeat this

164

labour so often until you have gotten enough ashes. These ashes of lead are fitted to receive an amendment by the spirits of salts, and afterwards to yield their gold and silver by fusion, and that with profit.

IV.   The manner of bettering the ashes of lead by the spirits of salts, and of extracting thencefrom the gold and silver with gain.

First of all, you must have plenty of the spirit of salt or AQUA REGIS, as concerning the easie getting such spirits, we have mentioned the way at large in the foregoing Appendix, and will yet treat more of them in this present Century.

Besides, you must also provide your self of red or reddish kind of flints, which (besides iron) do also contain in them a volatile gold. Out of these is the tincture to be extracted by the spirit of salt, or by AQUA REGIS; after that mariner I delivered at large in the first Century, and in the Appendix to the fifth part of the prosperity of GERMANY, and will yet farther teach more clearly and more compendiously in this Century.

These extractions are to be poured upon the lead ashes, that they may be well moistened therewithall; the unprofitable phlegm is to be evaporated by a gentle heat, and the fire to be augmented that the spirits also may follow; of which more heavy spirits there will be enough remaining in the secret Cementatory Pot, and as much as is sufficient for the due operation upon the lead, that so being bettered it may afterwards prove a gainfull emitter of its gold and silver. He that has a desire of exercising this labour with greater profit, may satisfie his desire, if he will but pour on such extractions twice or thrice upon the said lead ashes, that they may be concentrated by them afore they are cemented in the Cementatory Vessel, and may be reduced into the bettered lead. For by this means, all the labours and costs will be more largely recompenced, and the more plenty of gain gotten.

This now is the making the lead ashes, whether you do either per Se, or by the help of the other metals, convert it by Cementation into a better metal.

V. A brief description of the secret Cementatory Pot, which admits not of any spilling, and which is sealed with the Seal of HERMES, of which I made mention in the first Century.

Build with Stones or Clay or Potters earth such a Furnace as that is, which I described in the first part of my Furnaces, as necessary for the making of spirit of salt. But let the lower part thereof be a little broader that so the Metals being cast upon the Coals may not stick to the walls of the Furnace and so be somewhat lost, but may fall directly down on the live Coals. It must be made four square and of such a bigness as may serve the purpose according as you are minded to cement a greater or lesser quantity of metal therein.

VI. Of the Cover of the Cementatory-pot, what it ought to be, that so it may suffer nothing to get away in fume.

This Cover of this Cementatory Box or Furnace which I told you was to be made of the Lute of Wisdom, is not properly a Cover but a leaden Cistern, serving for the reception of those spirits which is to be filled with water, and 'tis to be so fitted to a pipe that is to come out of the Furnace, that the ascending spirits passing thereinto may be the better refrigerated by the water, and the sooner condensed, and saved for farther uses.

VII. Of the use and benefit of the secret Cementing Pot.

When any one has a mind to cement the Lead ashes, from which the extractions of the coloured Flints have been sometimes abstracted in the said Cementatory Box, and to graduate them, or so bring to pass that they may contain (or hold the) Gold and Silver, let him first of all fill his Furnace with Charcoal, and let him so order it that his fire may kindle by little and little till the Furnace be well heated red hot; till this is done, the Cover that is at top is to be taken so long off, that so neither the heat nor smoke may pass out at the side through the Pipe into the adjoined Leaden Cistern.

When the Furnace is throughly heated, and that 'tis now time to begin the Cementation, the top of the Furnace is to be shut with its Cover, that the heat may be forced to pass through the Pipe into the Receiver. Having so done, you are to fill an Iron Spoon or Ladle of your prepared Lead ashes, and put them into the Furnace at the fore hole which serves for the throwing in your Coals, the which ashes are to be so put in as to cover the Coals over, but not so as to choke them but that they may have air enough to burn, and that the fire be not put our, but do just in that manner as you are wont to distill the Spirit of Salt. By this means all the Spirits that remained yet behind in the Lead ashes, will betake themselves into the Receiver, and the Ashes of the Lead will be bettered by the graduating and tinging spirits, and will part of them be reduced into a body, and part will yet retain the form of ashes, and fall down through the Grate to the bottom of the Furnace. Then the Furnace is to be again filled with Coals, and more Ashes are to be put therein with a Spoon as afore, and this labour is to be continued so long till all the Ashes are consumed.

All the labour being finished, take out your Ashes together with the lead reduced into a body, melt them in the Furnace which is called STICHOFEN, they will melt wondrous easie, then put some small part thereof to the Test, thereby to try whether or no they are enriched enough, to be turned into a Litharge and undergoe the metallick separation.

If they won't as yet brook the trial, let the Lead be again turned into Ashes in your Iron Pot, & repeat the whole afore prescribed labour, and that so often till at length the Lead be rendred rich enough in Gold and Silver, the which may be converted into Litharge after the usual manner, and separated from the Gold and Silver. The Litharge being taken away, and gathered together, and broken in a Mill, serves for farther uses in this operation. The REGULUS of the Gold and Silver that is left upon the hearth is to be taken out and to be farther mundified in a Cupel after the accustomed way.

This is that more compendious incineration and reduction of Lead, which kind of bettering it, enricheth the operators with Gold and Silver.

N. B. That in this Cementation the sharp spirits do carry over with them some of the Volatile Lead into the Receiver, and there it settles to the bottom; the which powder being freed from all the Acrimony of the spirits by due washings, and being then dried, may be used to all such intents and operations to which the Mercury of Saturn is wont to be used, and which is made by dissolving the Lead in AQUA FORTIS, and precipitating it by Salt water.

N. B. This distilled Mercury hath more hidden under it than the other hath; for it carries hidden in it a Volatile Gold, which may be separated from it and improved about the gradation and Tincture of other Metals, and that with no small profit, concerning which we will say more afterwards.

Thus friendly Reader, hast thou my more compendious incineration and reduction into better Metals, the which I would not hide from thee, and hereby shall I satisfie those to whom the way prescribed in my Appendix is too tedious and laborious and they may make use of this way instead of that other, which withall is easier and will without all doubts yield more Gold and Silver than that other way.

VIII. Another emendation or bettering of Lead by the graduating extractions of coloured Flints.

Extract either coloured Flints, such as have in them Volatile Gold or Sand or Clay, by the spirit of Salt or AQUA REGIA, and draw off the Liquor by Distillation. If you thereto add Salt afore their extraction the dissolvent will receive encrease from the Salt, especially if done in such an instrument, in which a great quantity of extracted matters may be abstracted in a few hours, without either Cucurbits or the other commonly known distilling Vessels, and the same operation may be continued a long while. By this means, there is not only the least loss of your dissolvent, but it rather gets no small encrease from the Salt. By this instrument also, thou maist not only prepare great store of sharp spirits necessary for thy operation at the beginning, but likewise commodiously extract your Minerals, and separate the dissolvent again from the Minerals so extracted, so that you shall not lose the least particle of your dissolvent.

But forasmuch as all the Gold, Silver, and Copper may much easier be separated from its MENSTRUUM, by this so unheard of and never seen instrumtnt, than by the way of precipitation, 'tis altogether better and safer for a Man not to precipitate his extracted Metals, but rather draw off the MENSTRUUM from them, that so he may have them dry. And though that all the spirits go not wholly off, so as that nothing of them abide with the Metals, yet they do no hurt, but rather exalt the Litharge that

170

is put unto them into an higher degree, as it also does to the Ashes themselves of the Lead, when they are cemented together in the afore described cementing Furnace; in which Cementation the Volatile Gold is, together with the corporeal Gold conserved, and which otherwise would vanish away in the common melting Fire.

But if so be that any one has a mind to precipitate the Metals extracted out of poor Mines, after the manner prescribed in the Appendix, to the intent he may after the precipitation make Salt Peter of the remaining Waters, he may reduce the CALX'S of the Metals, and principally of the Gold very easily and without any loss, by this following way.

IX. The manner of reducing the precipitated and washed Calx of SOL without any loss.

The precipitation of Gold by LIXIVIUMS, Liquor of Flints, Spirit of Urine, Solution of Mercury, hath been clearly enough described in the Appendix to the fifth part of the prosperity of GERMANY; but yet the reduction of the same was past over in silence, because of the too much haste of the said Book. Therefore it seemed unto me necessary to insert the same here, for their sakes, who have but little knowledge, or in a manner none at all in these affairs; for should I go to propound such a thing for the skillfull Chymists, I should but do what is already done, especially because he deserves not the name of a Chymist who is ignorant of the reduction of the solar CALX.

But forasmuch as it may so happen, that even the

unsklllfull may set about this extracting of the Minerals, and yet be ignorant of the way of reducing the Gold though they should have extracted it; therefore have I judged it not amiss to illustrate that reduction by my describing thereof here, the which being divers, according as the precipitation is made by such or such a means, doth also require different operations.

X. The reduction of the solar CALX precipitated by the Liquor of Flints.

Albeit that Borax reduceth every CALX of SOL to its former body, if it be therewith mixed and melted in a Crucible, yet that would prove too dear, if somewhat a greater quantity thereof be required for the reduction; for there must be of it at least twice or thrice as much in weight as is of the Gold, if you would have a due reduction of the Gold made. The reason is this, because the Flints precipitated to the bottom together with the Gold, and so sticking on to the Gold impead its fusion so that it cannot rightly come together into its due body. Hereupon is it necessary that there should be the double or treple weight of Borax added to the Gold if you would have all your Gold return unto its former body without detriment.

But whereas there are also other matters to be found which make the Gold fusible and are not so dear as Borax is, the use of such things is to be admitted, but especially when a great quantity of Gold is to be reduced. Otherwise if it be but little Gold that is reduced, and you have not the aforesaid matters at hand, one may for such a small trial use Borax. But where there is a greater quantity to be reduced the following matter will be found to be far more profitable and beneficial.

XI. How the Gold which is precipitated by the Liquor of Flints, is to be melted without Borax, by the Glass of Lead only, which is of a far meaner price.

Take of your Gold precipitated by the Liquor of Flints and dried, one part, of Glass made of Lead and beaten into powder, three parts; the which mix well with the Gold and put into a Crucible, which said Crucible let be put into another bigger one (for which operation the Hassion pots are most fit) that so if the Gold chance to flow out of the inner pot, it may stay in the outer and be conserved. For the Glass of Lead is of such a nature that it usually perforates or runs through the pot. Having thus done put your twofold Crucible containing your coininixed matters into a wind Furnace, such an one as I have described, and when you have covered it, put Coals under it (or about it) and urge your fire for one quarter of an hour, that all may well flow, then pour it out, and separate the REGULUS of Gold with a stroke or two from the glass of the Lead; which said Glass hath attracted to it self all the flinty matter, and suffers the pure and malleable Gold to settle to the bottom into a REGULUS.

N. B. If your glass of Lead be still yellow as it was before the operation, 'tis a sign that all the Gold is separated therefrom; but if it be of a green colour;

tis a sure sign that it has as yet some Gold mixed with
it. For Gold being mixed with Glass shews its being
there by yielding a skie-colour, the which skie-colour
is necessarily changed in the yellow glass of Lead into
a green; because every yellow and skie-colour do in
their commixtion beget a green.

Now then that you may get out the reliques of the
Gold out of the leaden Glass you must proceed the
following way.

XII. By what means the Glass of Lead which as yet
contains in it some reliques of Gold is to be dealt
withall, that it may let them go out of its body.

Melt that Glass of Lead in a well covered pot,
that I mean in which you suppose some Gold to be, and
being well molten cast in a little iron filings, and mix
it well by stirring it with an iron rod, and leave it in
the fire thus molten, for one quarter of an hour, that
the sulphur of the Glass may be killed by its corroding
of the iron, and may let fall a leady REGULUS wherein
the Gold will be, and which (in the first melting) the
Glass held up, will separate it by the Cupel, from the
Lead.

N.B. But here you are to observe that the filings
of the iron are to be used sparingly to this
precipitation; for by how much the more iron is added,
so much the greater will the REGULUS of the lead be, and
consequently require a greater Cupel, which is not so
necessary.

For put case the Glass of Lead in which the Gold

is suspected to be is about one pound weight; and there is but about a QUINTA or certain small weight of Gold; now it is not necessary to have any more than one Lot of Lead or thereabouts, precipitated thereout of into a REGULUS, to which precipitation is required no more than one Lot of the filings of Iron. For the REGULUS of Lead precipitated out of the Glass, doth for the most part answer in weight, to the weight of the Iron filings used about the precipitation, or to speak more clearly, you will get so much leaden REGULUS, as the Iron is you added.

The remaining Glass becomes black and is unprofitable for any farther melting with Gold, but yet needs not be cast away, because those SCORIA'S do yet contain much Lead, and therefore serve to be mixt with such Pots as you have used and broken about Metals, or with other wild and hardly fusile metallick Veins, to render them fusible, being I say commixt with these, and put in the Furnace which the GERMANS call STICHOFEN, do not only yield forth all their Lead, but withall draw out the Metals out of those matters which were mixed with them in the melting. But they are principally profitable for the melting and reducing of those Metals, which do not only very difficulty admit of fusion by themselves, but withall do, being mixed with the Ashes of Tin, so much the more difficulty suffer themselves to be reduced by melting, unto their former bodies. But in defect of such Metals and Minerals, as are not but with much adoe tamed by Liquefaction, you may put to that black Glass of Lead, one fourth part only of filings, or

SCORIA'S of Iron made into powder, that so both the matters thus commixt may be molten in the Furnace STICHOFEN. So by the addition and help of the Iron, all the Lead will be reduced to its former body, and will withall extract out of the Iron whatsoever of Gold and Silver lay therein hidden; so that by this means there may be reaped a great benefit from this reduction of the Glass of Lead. But yet that Lead is to be tried by a foregoing tryal, whether or no it be rich enough in Gold and Silver to quit the costs of separation? For if it be not, it must be used to the afore described incineration, that so there may be no loss either of the Gold or the Lead.

XIII. The preparation of the Glass of Lead, for the reducing such Gold as being precipitated by the Liquor of Flints, is of difficult fusion.

Take of white and fusile Flints (or Pebbles) one part, and of MINIUM, or any other Ashes of Lead, or else even of Litharge it self four parts, each of which being powdered apart, you are to commix and melt them well in a strong double Pot, then pour them out, and you will have a Hyacinth-coloured Glass, the which Glass is to be powdered and mixt with the Gold, and it makes the Gold Powder which resisteth melting fusible.

XIV. Another way of reducing Gold precipitated by the Liquor of Flints.

To one part of this hardly-melting Gold which is precipitated by the Liquor of Flints, admix two or three parts of Litharge, which matters put in a strong double Pot, and cover it well, and melt them well down in a Wind-Furnace, that the Litharge may draw unto it self all the Flints, and all the Gold may separate. Having separated the REGULUS from the SCORIA'S of the Lead, you must precipitate these SCORIA'S, which do as yet hold in them some small portion of Gold into a small leaden REGULUS, with the filings of Iron, as we shewed you but now, that so you may also have even that residue of Gold. The SCORIA are conserved by being reduced in the Furnace STICHOFEN, according to the operation already spoken of.

XV. Another way of rendering the Gold precipitated by the Liquor of Flints fusible.

Take of the said Gold one part, and the fixt Salt made of Salt Peter and Tartar, by combustion or calcination, three parts; commix them and melt them down in a crucible well covered. In this comelting the Salt swallows up the Flints, and the Gold being at liberty settles to the bottom. Pour out the molten mass, and separate the REGULUS of the Gold from the Salt, the which being dissolved with common water gives you your Liquor of Flints, to be again used to precipitate more extracted Gold.

This Salt doth not so easily perforate the Crucibles as those Glasses of Lead do, and therefore is it to be accounted of as the best and easiest of all these three prescribed ways.

XVI. The way of reducing Gold, precipitated by the Spirit of Urine.

The Spirit of Urine or of SAL ARMONIACK doth perfectly precipitate all the Gold out of the AQUA REGIS; the which being washed and dried, doth not admit of reduction after the manner of the other Gold, for if it be but only heated a little before it becomes red hot, it presently takes fire, and fulminates with a far more dreadfull noise than any Gun-powder. For if you put a small portion of the same, and no bigger than a Pea in a Silver, Iron, or Copper Spoon, and put it on the Coals that it may wax hot, it will give such a horrible crack, that 'twill even dull the hearing, and make a dent in the Spoon as if it had been beaten in with a Hammer. From whence it may easily be conjectured, that if somewhat a bigger quantity be put in a Pot on the Fire, it would make Pot and Furnace flie, by its so dreadfull thundring a stroak into most small shivers.

So then there is need of great wariness, to prevent the happening of so great danger, which is easily prevented by the following manner of operating.

XVII. By what means the fulminating force of Gold precipitated by a LIXIVIUM, or spirit of Urine is to be taken away.

Mix with this Gold precipitated by a LIXIVIUM, or by the spirit of Urine, half a part of Sulphur reduced into Powder, and let the said Sulphur be removed therefrom by burning amidst live Coals; for so being despoiled of that fulminating force, it may without danger be reduced by any kind of such matters as promote fusibility.

XVIII. By what means Gold that is despoiled of its fulminating force, by means of Sulphur may be reduced.

      Forasmuch as this Gold is void of all impurity, there needs (not) the addition of such matters as promote fusion, seeing it is of it self prone enough to melt. But yet least some grains of the Gold should stick on to the Pot, 'tis expedient to add some portion at least of such a kind of matter as accelerates or hasteneth fusion. And for this work, Borax, and the dry Liquor of Flints are excellent, of which if you add but one half part only to such Gold, (or, if you take of the Flints prepared with Salt of Tartar) it will by that means presently melt, and the Borax, or Liquor of Flints will not retain the least doit of the Gold.

XIX. The mariner of reducing the Metals that are not gotten out of the Waters by precipitation, but are freed from them by abstracting them.

      The Metals which are extracted out of the Mines, and freed from the waters by the abstracting of the dissolvent, cannot be so pure as those are which settle to the bottom by precipitation. For it is very rare for Gold and Silver to be found in metalline Veins, Stones, or Clay, without being commixed with other Metals; because for the most part, Copper is mixt with Silver, and Copper or Iron with Gold, the which being unseparated in the reduction makes the Gold and Silver impure. But now in the precipitation one Metal is freed after another from the MENSTRUTJJVI, and are not mixed with each other. But on the contrary, in the way of abstracting it, all the metals remain mixt together without any separation, and require a new separation and consequently a double labour, and more expences.

This inconveniency may be easily remedied by him who is versed in the knowledge of my dry separation of Metals. I have mentioned it in divers places of my writings, so that it would be needless to trouble the Reader with a superfluous rehearsal of the same in this place.

But forasmuch as every one hath not by him all my writings, I believed that it would be worth while, if I should here set down that Layer or Bath which washeth off the Metals with the help of Salt peter by the dry way. For, without the knowledge of this Artifice of separating the extracted Metals from each other, there would be yet requisite much labour, and much costs for the obtaining of the said Metals. But they are very easily, and with little labour, and with small costs separated the one from the other by the way here by us described, and indeed with more gain than is wont to be had by the way of precipitation.

And even as in the precipitation of Metals there is always some (portion) of the Waters, that puts on the nature of Salt Peter, viz, when the Waters that have been used, which are as it were the Seeds of Salt Peter, are implanted in an Alkalisate Salt, and so do multiply themselves in a wonderfull manner.

So likewise in the dry separation of Metals, there is in a manner, yet more Salt Peter gotten, viz, thus when they are separated in the melting Pots, from each other by Salt Peter, and by an artifical precipitation of one Metal after another, the Salt Peter you used is rendred fixt and Alkalizated, which Alcalizated niter is to be accounted of, as the root of Salt Peter. This root being implanted in acid Salts, is in like manner enriched with a plenteous encrease, and reduced into natural and inflamable Salt Peter; for, by it do the sharp Waters get to themselves the nature of Salt Peter, from those Alkalizate Salts. And if so be you seek not after the common Salt Peter, it is better to sow the Seed of Salt Peter (that is, some spirit of niter which you have used) into the appropriate root of Salt Peter, that is, into fixt niter. For by that means you will have (at the encrease) a wonderfull Salt Peter, which, in all operations, doth far more powerfully act than the common Salt Peter, what way soever it be mundified by; which is evidenced in my foregoing first Century.

Therefore forasmuch as in the separation of Metals by the dry way, there remains (after the operation is over) so much fixt Salt Peter as there was of nitrous Water in the moist extraction, it always abundantly supplies both Seed and Root of Salt Peter, so that they may be exceeding plentifully multiplied by other Salts, nor will you have any need of buying any more new Salt Peter, for the now spoken of Labour. Verily this is a most compendious way, not only of separating all Metals even in fusion, but also of somewhat bettering them, as shall be afterwards demonstrated.

XX. By what means such Gold as is commixt with Iron or Copper, and from which (being extracted out of the Minerals) the dissolvent has been drawn off, is to be reduced.

Let such unclean Gold be commixed with two or three parts of its weight of the Glass of Lead, and melt them in a strong Crucible. If there happen to be much Iron, it will of its own accord yield a leaden REGULUS, which being forced by the heat of the Fire in a Cupel will leave your Gold pure, because the Glass of Lead is wont to attract unto it self Iron and Copper. But if so be there is but little Iron mixed with your Gold the REGULUS of Lead will not separate or precipitate in the melting, and therefore as it melts some filings of Iron are to be added, and to be accureately stirred with a red hot Iron, that so a REGULUS of Lead may fall to the bottom, bigger or lesser according to the muchness or littleness of the Iron you added.

XXI. Another proper and fitting matter to reduce such Gold as hath Iron in it.

Take of Salt Peter one part, and of Antimony four parts, reduce them into a black Glass, by melting them. This Glass being powdered and eommixt with a wild or raw and not fusile Gold and so molten, precipitates the REGULUS of the Gold to the bottom, and brings the Iron into SCORIA'S.

XXII. The separation of the Antimony from the Gold.

Such golden REGULUS'S do not admit of separation in the Cupel, like as those do which the Glass of Lead is used to. Therefore Salt Peter is to be used in the melting Pots or Crucibles, to make the separation of them.

Put the Antimonial REGULUS in a melting Pot, melt it down in a Wind Furnace, and being molten cast in by little and little some dry Salt Peter, that so it may seize upon the REGULUS and transmute it into SCORIA'S. The SCORIA flowing in the Pot like water, are a sign that the Gold is well cleaned, and that all the Antimony is reduced into SCORIA'S. Then pour it forth into a Cone that it may cool, and the pure and malleable Gold will settle into a REGULUS at the bottom. Now all the Salt Peter is rendred fixt in this operation, then if you put your SCORIA'S again in the Crucible, and put into it some Coals and melt your SCORIA down, almost all the Antimony being freed from the Salt Peter will gather into a REGULUS, and will again serve for reducing of more Gold; for it will as readily reduce your extracted Gold unto its former body, as the (aforesaid) Glass it self will. But this labour requires a diligent Operator who knows how to handle it with singular skill, though it be easie, and requireth not any great Artifice, but only an accurate diligence, which use only makes a Man skilled in.

The Salt Peter used about this labour, gets the nature of an Alkali Salt, and being put on the live Coals doth no more burn, but being dissolved in Water

yields a sharp LIXIVIUM, very proper for many
operations, and serving instead of a Lye made with Wood-
ashes. But the chiefest use thereof is this, viz, seeing
it is the true root of Salt Peter, it may be added to
other Salts, out of which in process of time, it will be
notably augmented and produce new burning Salt Peter. He
that has a desire gainfully to augment this fixt Salt
Peter with common Kitchin Salt, and again to transmute
it into inflamable Salt Peter, may accomplish his desire
if he makes use of the following operation.

XXIII. The way of making most excellent and inflamable
Salt Peter in plenty, and with profit out of common
Kitchin Salt and the LIXIVIUM of Salt Peter that has
been used.

There is so small a difference betwixt common
Kitchin Salt, and Salt Peter, that the Salt may easily
be turned into Salt Peter, and that by several
operations, as well by the seed of Salt Peter as by
sharp spirits, as we have taught above, or even by fixt
niter which operation we will here shew.

We will use for an example, the baking of Bread,
and the brewing of Ale. If when the Meal is with Water
brought into Dough, there be added unto it but a few
grounds of Ale or Leven, the whole mass begins to heave
it self up, and becoming thin (or light) is rendred fit
to be baked into Bread, the which hath altogether the
same property as those few Ale Grounds, or that little
Leven had. And so that very self same Dough is likewise
fit to make other Meal ferment, even to infinity. The

184

same likewise observable in the brewing of Ale, so that he who hath but once only so much Ale Grounds (or Yest) or leavened Dough as served his turn once, may brew Ale and bake Bread even to infinity. So likewise is the same evidently manifest by the encrease of Vegetables, which may be infinitely multiplied by the Alkalized Salt of the Earth, if you have but once their Seeds and Roots. In like manner may the same propagation be performed by another way, viz, by ingrafting of that which you would propagate into another of the same kind. For example, I have in my Garden excellent Apples, Pears, Cherries, or such like Tree-fruits, and I have a mind to see more of them in my Garden; there fore do I cut off some branch, or perhaps even the Tree it self to the trunk or body, of some wild, or at least not so noble a Fruit-bearing Pear Tree or Apple Tree, and therein, viz, in that branch or stock, do I ingraft according to Art some little boughs or cions of some other Tree that bears excellent Fruit, and which I desire to encrease, the which Tree now doth no more produce the wild and degenerate or bad Fruits, it did according to its kind, but such Fruits as the Tree whence the cion was taken, bears.

By these kind of similitudes may any one that hath understanding easily see, that it is possible by Art, to transmute one nature into another, if, viz, the Seeds and Roots of things are applied to this transmutation. But now if any one should plant a stalk or leaf in the digged earth, and would thereby encrease or propagate it, he will never see any success of his labour; for the

185

stalks and leaves would rot and so no new Herb would again bud out from them as is wont to be out of the Seed and Roots.

Even on this wise is it with Salt Peter, which if it be mixed with common Salt it would not verily produce any encrease, as 'tis wont to do out of its Seed and Root, as we have already laid open.

Such likewise is the nature of Metals, touching the propagation and encrease of which their proper seeds and roots are requisite. What I pray are those Tinctures, (one only particle of which and that no bigger than a Pea, being cast on an whole pound of Tin or Lead, transmuteth that same Metal into pure good Gold, and changeth and augmenteth itself (as being the true seed of Metals) a thousand fold, out of so gross and earthly a body into so noble and so golden a nature in so short a space of time, what (I say) are those Tinctures, but the very seeds of Metals, and the very metallick roots. But by what means they are to be obtained, and to be brought under a Man's power, for my part I do not know. But yet I could not but deliver my simple opinion and conceptions concerning this thing, to the studious of Art.

'Tis certain that all Metals have their rise out of one and the same Seed, but that they differ so very much amongst themselves, and that one becomes more ripe than another, is to be imputed to the diversity of accidents. In one and the same Tree are produced Blossoms, and small Fruits of an unpleasant taste; then afterwards bitter and immature ones, and at length ripe

186

and sweet ones, and are not alike either in form, odour, or savour, nor are they of like effects, and yet do they all arise out of one beginning, viz, out of their Seed and Root. So is it even with Metals.

For as touching their Seed I do verily believe, that if from the most soft and as yet most immature Metals, such as ZINK, Lead, Tin, Antimony, Bismuth, Cobolt, & etc. their stinking combustible and superfluous Sulphur, could by some Bath or other be so taken away, as that nothing may remain save only a most pure Mercury, that then I say such a Mercury, or such a Seed of Metals may be easily transmuted by pure Gold, as being the most pure Root of Metals into a true Tincture.

But to turn common Salt into Salt Peter, the operation is thus. Take of one part of black or of any other common Salt, and mix it with two or three parts of CALX VIVE being reduced into Powder by lying in the Air, and lay it in such a place as lies open to the Air and Sunbeams, but yet keeps off the Rain, as we have taught in the Appendix.

Moisten this heap with the abovesaid LIXIVIUM of Salt Peter, and being dried, repeat the moistening and drying so long, until the ferment shall have converted all the common Salt and turned it into inflamable Salt Peter, the which doth either sooner or later happen, according as the operation hath been the more diligently or negligently handled. All being turned into Salt Peter, let an extraction be made with common Water, as the usual custom is, and lay the Reliques in the aforesaid place, and again moisten them with the said

LIXIVIUM as afore, or in defect thereof sprinkle them with common Water, still moistening them after each drying, until there be a new encrease of Salt Peter be gotten, the which you are to wash off with common Water. And so this operation proceeds, or holds on even to infinity.

XXIV. Another far more compendious way of converting common Salt, by the help of fixt Salt Peter into excellent Salt Peter.

Mix some certain weight of common Salt dissolved PER SE in common Water, and as much of fixt Salt Peter likewise dissolved in common Water, mix them in a wooden Vessel; in which Vessel the fixt Salt Peter being as it were a ferment will seize upon the common Salt, and turn it by fermentation into excellent Salt Peter.

He that disires a more mature Salt Peter may instead of the solution of common Salt, pour upon the fixt Salt Peter LIXIVIUM, those sharp waters of Salt Peter, which have already been used about other labours, and they will seize upon that LIXIVIUM with a more vehement operation, so that of both the solutions as well the acid and spiritual, as the fixt and corporeal Peter, there will be gotten in a few hours space, the most excellent Salt Peter and such as cannot be by any other way whatsoever purchased.

N. B. If any one has a mind of getting a greater quantity of Salt Peter, he may first dissolve his common Salt in the sharp Water of Salt Peter, and (mix it) in that self same LIXIVIUM, (viz. of Peter) and after the mixing of these two contrary solutions evaporate the common Water, that the Salt Peter may shoot into Crystals, of which, there will indeed be a greater quantity, but then it will not be so good as that which was made by the first operation.

XXV. Another gainfull way of making good and burning Salt Peter out of common Salt, by the help of fixt Salt Peter.

Commix equal parts of the SCORIA'S of fixt Salt Peter that you have used, and of the common Kitchin Salt together, and add thereunto twice as much CALX-VIVE first reduced into Powder by lying in the Air, (as they both weigh). Of this mixture make up round Balls, and so pile or stow them with Wood, that it may be a STRATUM SUPERSTRATUM, (or a Bed of each orderly[3]) as the Chymists call it. Kindle your pile of Wood and let all your Balls be red hot for an hour:

And the fixt Salt Peter will by a wonderfull inversion transmute the nature of the common Salt, and turn it into Salt Peter, but yet not inflamable till the Salts have been moistened some due time, and so attracted a life out of the Air, and made fit to conceive a flame, or to burn.

N. B. If instead of Rain water you use such Waters as you have already used and extracted your Minerals withall, to moisten your mass with, then will you thence get in some few Weeks space, an inflamable Salt Peter.

But forasmuch as in the extraction of Minerals and

---

[3] Layer upon layer.

separation of Metals, there will be such a great quantity of sharp nitrous Waters, and likewise of fixt Salt Peter offer themselves for the accomplishment of this operation, and so great a benefit and gain is gotten by that so plentifull an augmentation of your Salt Peter (which hath already sufficiently profitably paid your costs) out of vile and common Salt; hence clearly follows, that all those hitherto described labours and operations are effected, in a manner without any costs or expences, which is indeed an unheard of thing, but yet most true, and exceeds the belief of ignorant Men.

XXVI. The reduction of Silver extracted out of the Minerals, and freed from the AQUA FORTIS by abstraction, (or drawing off the said AQUA FORTIS).

The Silver from which AQUA FORTIS has been drawn by Distillation, needs not any matter to help on fusibility, for as much as it doth of its own nature admit of a very easie Flux; but that the Fugitive Spirits that adhere unto it would carry away somewhat of the same. So now, to prevent this discommodity, you may add unto such a Silver a little of the fixt Salt separated out of the LIXIVIUM (of fixt Peter) the which Salt Alkaly will mortifie the acid spirits so, that they shall not be able to carry off any thing at all in the melting.

XXVII. The reduction of extracted Copper.

If the Copper be not mixt with any other Metals, and be but little in quantity, it may be reduced so in Crucibles by it self, but if it be in a plentifull quantity it may be done by blast.

But where if contains Iron or LAPIS CALAMINARIS, (which two the Minerals (of Copper) do frequently abound withall) there it admits not of reduction per se without the help of other matters, because of the Iron, ZINK, or CALAMINARIS; which Minerals associating themselves with the Copper, in the melting are wont to make it brittle. But this inconveniency may be prevented the following way.

XXVIII. The way of making Copper, which hath Iron in it malleable by reduction.

Mix such Copper as hath in it LAPIS CALAMINARIS or Iron, with common Salt, and put it in a Crucible and melt it, that so the Salt may associate or draw unto it self the Iron or LAPIS CALAMINARIS out of the Copper, and turn them into SCORIA'S, leaving the Copper, which will settle to the bottom and go into a REGULUS.

XXIX. By what means Copper is to be separated from the Silver, if they are both together extracted out of the Mines, and the Silver has not been precipitated out of the solution by the Water of Salt, but the dissolving MENSTRUUM hath been abstracted from them so conjoined both together.

If the Silver be more in quantity than the Copper, then the Copper is easily extracted out of the other by

191

the Water of Salt wherein a little Tartar hath been dissolved. For Salt and Tartar do readily dissolve Copper, and leave the Silver.

But if the Copper bear the Bell, and there be more of that than the Silver, then will it be better to precipitate the Silver first by the Water of Salt, out of the first extraction of the Minerals; and afterwards the Copper will be likewise freed by abstracting the dissolving MENSTRUUM, insomuch that each of these two Metals are gotten apart.

XXX. If the extracted Copper comprehends in it any Gold, by what means the Gold may be therefrom separated.

Albeit if a solution of SATURN or LUNE being poured on the dissolved Copper, and well shook with the same solution will fish out some Gold, yet it gets not all unless it be debilitated by some LIXIVIUM. But now the LIXIVIUM being poured thereunto that so the solvent being debilitated may the easier let go its Gold thereby, hath with it this inconveniency, viz, that the solvent is made wholly unprofitable to be used about any more extractions. Nay more, there's also this discommodity, that if an error be committed by pouring on a little more LIXIVIUM than is expedient, there will also precipitate some of the Copper together with the Gold.

To prevent therefore these inconveniencies, the solution of the Copper which contains in it Gold, is to be drawn off even unto driness, in my secret and by me invented distilling Vessel, in which Vessel it may easily and in great plenty be done, and the following MENSTRUUM which dissolveth only Copper and not Gold is to be poured upon the dried matter, that the Copper may be dissolved, and the Gold be it either much or little may remain in the bottom undissolved. The dissolved Copper may be precipitated out of the Water with a LIXIVIUM, whereto is added some of the Liquor of Flints, and be washt and dried, and with strong Vinegar be turned into a most delicate Verdigreace. The Water that is thus made use of, if it be poured on Alkalizated Salts yields good Salt Peter.

He that does not much regard that green colour may separate the dissolvent from the Copper by Distillation, and again use the same for the like dissolving of new Copper.

But now there must be in the Copper so much Gold as to quit the costs of this labour, and to prove gainfull; otherwise it is better to leave the Gold with the Copper than to buy it at so dear a rate.

XXXI. The making of such a MENSTRUUM as dissolveth the Copper and drives from it self, or precipitates the Gold.

This dissolvent is no other than AQUA FORTIS, wherein a little Tartar is dissolved. For the Tartar being an enemy to the Gold, is wont to precipitate the Gold out of the solutions like as common Salt doth Silver out of AQUA FORTIS. By this way may all the Copper be easily separated from the Gold, concerning which, more shall be spoken in its due place.

XXXII. Another reduction of Copper that hath Gold in it, and the perfect separation of the Copper from the Gold.

Add to such Copper that hath Gold in it, some Silver, and melt it with so much REGULUS of Antimony as is twice the weight of the said Metals thus together taken. Separate the said REGULUS together with the Copper from the Silver by the addition of Salt Peter, that so the Silver may retain with the Gold that was in the Copper, the which is to be afterwards separated if it be worth the while. Now it is not necessary presently to separate it, for asmuch as it is far better many times to abstract such a Copper that has Gold therein from such a Silver, that so the Silver may be enriched with a great quantity of Gold by so many abstractions; the which abstractions require but very little costs besides the charges of the Fire and Crucibles.

For all the Salt Peter used hereabouts, together with all the Copper and all the REGULUS of Antimony may be thencefrom again recovered, by him who rightly knows the precipitation. Besides, there lies hid under this operation, some great matter as concerning the amending of the Metals. For it is a way of arriving to the know-ledge of impregnating all Silver by Copper, with Gold, and Copper it self with Silver; concerning which thing there are more instructions to be found in other parts or places of my Writings.

These things may at present suffice, touching the reduction of the Metals extracted out of the poor Mines, and (concerning those things which by reason of the hasty Edition of the Appendix to the fifth part of the prosperity of GERMANY were omitted therein) the which defects the well minded Reader may from hence supply.

XXXIII. A brief description of the above mentioned artifical Instrument, by the help whereof the spirits necessary for the extraction of the Metals out of the poor Mines that contain in them Gold, Silver and Copper, are plentifully prepared, the Minerals themselves extracted, and the dissolving MENSTRUUMS, again easily separated from the Metals.

This incomparable and by me newly found out Instrument, being most profitable and commodious for the easie extractions of Metals, and preparations of the dissolving MENSTRUUMS, is made of a peculiar earth, and is almost of the figure or likeness of a Bakers Oven, and is either of a bigger or lesser size, according to the quantity any one has a mind to labour in; In the forepart it hath a Door, and in the end (or top) or very near it, it hath an Outlet. To the Cover serving instead of an Alembick, a great receiving Vessel is to be fitted, fit for the reception of the outgoing spirits. After that the Furnace is heated, the prepared Salts being put in peculiar Pots or Crucibles made of the best earth are to be put with a pair of Tongs prepared for this peculiar use into the Instrument, and all the spirits will be drawn off with a speedy Distillation. Now there is no danger here of breaking the Instruments, and the Distillation may be done in the space of one or two hours, how great a quantity soever of Salt was used to the Distillation. When the Distillation is over, the Pots that were put into that Instrument or Furnace are to be again taken out with your Tongs, and presently other Pots filled with Salts are to be put in the room of them you took out, and the spirits again driven out by a new Distillation. This labour may be kept on as long as one pleaseth, or as long as he hath any matter to distill withall; because the Vessel never cools as long as the Distillation is continued. This Furnace therefore is most notable fit for the Distillation of a great quantity of Salts, and that by a labour which is so exceeding speedily finished.

The same way of Distillation is to be observed in the extraction of Minerals or Metalline Earths, the which can be far sooner extracted and far speedier this way, than by that described in the Appendix which is to be done by heating the Glasses.

After the same manner is the dissolving MENSTRUUM it self speedily again abstracted from the extracted Metals, and being thereby preserved without any loss is to be applied to farther use. This Instrument therefore doth so compendiously and easily dispatch all those said labours, that (set aside your Fire and Salt) the plentifull making your spirits, the abundant extraction of the Minerals, and the separation of your MENSTRUUM'S from the extracted Metals and its preservation, are in a manner done without any costs.

XXXIV. Now follows an explication of some secrets effected by the help of my SAL MIRABILIS, concerning

which there is mention made in the second part of
MIRACULUM MUNDI.

It is clearly evident from many places of my
Writings, and principally in the second part of
MIRACULUM MUNDI that my SAL MIRABILIS is diversly
prepared; hence it follows of course, that the use
thereof is different. For it hath one use when (after
the spirit is thencefrom distilled) it is taken out of
the Cucurbit, and hath as yet a corrosive nature.
Contrarily, it hath another kind of use when this
corrosive Salt is dissolved in common Water, and
filtred, and set in the cold, that so the best part
thereof may shoot into long Crystals, which having no
corrosive power, serve for a peculiar use. It hath
likewise another use when it is deprived of all corros-
Mty and turned into a sweetness, as I have shown in many
places of my Writings. This is to be known by such as
would use it, for this or that labour, that so they may
commit no error, but be thereby rendred Masters of their
desires the more easily.

We will therefore make inspection into some of
those principal secrets which are declared in the second
part of MIRACULUM MUNDI, and examine whether or no they
can be effected after the same manner I prescribed?

XXXV. By what means any Water, Wine, Ale, Vinegar and
other liquors may be coagulated in a few hours space
into hard pieces like ice, by the SAL MIRABILIS.

For such a coagulation of all watery and moist
things, well edulcorated (as the Chymists phrase it) SAL

197

MIRABILIS is to be taken and such as is shot into long Crystals, prepared of an equal weight of Salt and good Oil of Vitriol, because a most great driness ariseth from the Oil of Vitriol.

Such an excellently well prepared SAL MIRABILIS, and which is shot into long Crystals, is to be reduced (by calcination in the Sun) into a fine powder, that so it may lose all its moistness and yet not melt. For if it melts, then it would need grinding again; one part of this calcined SAL MIRABILIS is able to coagulate three parts of Water, Wine, Ale, or any other liquor which it is mixed withall, into a dry matter like to Ice, insomuch that it may be carried in a Sack or a Seive full of holes.

But what use such a coagulation may serve for, would be too tedious to declare in this place. Ar.y one will find what use is to be thereof made, if he well meditates upon the thing.

XXXVI. The separation of the Water, Wine, or Ale, from the SAL MIRABILIS.

The coagulated liquors may be commodiously separated from the SAL MIRABILIS by Distillation; but the aquosity of the coagulated Wine and Ale are to be separated only by Distillation, and the grosser part remains behind in the Cucurbit with the Salt. But the SAL MIRABILIS is by calcination, again freed from all impurity, and again made white and fit for any other such like new effect.

N. B. I doubt not but that there are other ways of

coagulating watery liquors into Ice, concerning which we shall say somewhat in their due place.

XXXVII. How the sharp spirits of Salts, or AQUA FORTIS, AQUA REGIS, Spirit of Salt, Spirit of Vitriol, of Allum, and the like may be coagulated into hard Salts, not unlike to frozen Water.

This coagulation of sharp Spirits out of Salts, is done the same way as the coagulation of common Water, and other sweet liquors is performed by; but the separation ought to be done in Vessels of the best Earth, or in Glass, because of their sharpness. And certain it is, that with these coagulated Spirits of Salts many things of great moment may be done, the mentioning whereof we for brevity sake do here pass over.

For I have purposed to demonstrate at this time, some secrets only which are mentioned in the second part of MIRACULUM MUNDI, and to assert the truth of them.

By these two described coagulations any one may easily learn that the coagulation of other moist things are possible to be done.

XXXVIII.　How the head of a fountain may be stopped up by this SAL MIRABILIS.

It sometimes happens that there breaks out a Spring of Waters in some places where it proves offensive and hurtfull. And forasmuch as they are sometimes very difficult to be stopt up, I will set down a way in this place of stopping it by SAL MIRABILIS, but

chiefly to this end, that the nature and property of
things may be throughly learned, and besides, that even
Arts and Sciences themselves do sometime bring no small
help, especially when no counsel avails. Take therefore
of your SAL MIRABILIS, heated red hot as much as is
sufficient, wrap it up in a linnen cloth and thrust it
into the hole of the Fountain, and it will be turned
with the Water into an hard Stone, and thereby enforceth
the Fountain to seek it self some other passage.

XXXIX. The way of separating the Phlegm from subtile
Spirits.

    Because the Volatile and sulphureous Spirits of
Salts are of great efficacy in Medicine, and principally
when their Phlegm or aqueous humidities are removed from
them, the which thing every one can't bring to pass. I
have therefore judged it worth while, even for the sake
of the Sick, to discover an easie way of so doing, by my
SAL MIRABILIS as follows.

    Fill a Glass Cucurbit half full with SAL
MIRABILIS, pour thereupon the Volatile Spirit of
Vitriol, Niter, or common Salt, and distill thence by B.
the most subtile Spirit, the which will come off, and
leave the unprofitable Phlegm behind with the SAL
MIRABILIS, the which (by heating red hot) you may again
render fit for new operations.

XL. Another and easier way, yea even almost an
incredible and miraculous one of freeing Wine, Ale,
Vinegar, Brandy, and all other moist liquors from their

uriprofitable Phlegm in a moment of time, by my SAL MIRABILIS.

The precedent coagulation of moist liquors ariseth from that most great driness which lies hid in the SAL MIRABILIS. But this way we now deliver, proceedeth from the concentrated cold of moist Fires, which Fires we have treated of in the first Century, and 'tis thus.

Take one pound of the abovesaid SAL MIRABILIS, put it in a strong Glass, and pour thereupon two parts or pounds of the concentrated and cold Fire of some Salts, whether it be of Vitriol, or common Salt, or Salt Peter, whose Fire excells the Fires of other Salts; and let them lie quiet for some hours, and there will be made an Icy mass of them both, the which you shall in the Winter time set out in the Snow or in some cold place which by how much the colder so much the better; where the longer it abides in the cold, the more will the cold Fire be concentrated, and consequently so much the greater matters may by such a concentrated body be effected,

XLI. The receiving or catching the breath of Men, as they sit in some warm Stove, and the changing it into the form of Ice.

If thou hast a mind to create a kind of admiration amongst thy Guests or Friends when they are with thee, and to give them some profitable recreation, you may accomplish your desire the following way.

Carry with thee a Glass full of the moist Fire of Salt, and which is coagulated by the SAL MIRABILIS, and hath stood some hours in the cold, into the warm Stove, and hang it up over the Table by a thread or small line, when your Guests are set at the Table, and when they ask you what this signifies, you may tell them that you will for their Recreations sake, shew them some pleasant diversion; after they have made an end of eating and drinking. Upon this they will all of them have a desire to see those tricks and ever now and then cast up their eyes upon the Glass. But after that the Glass has there hanged a minute or half minutes space, the breath of the Men that sit about it will presently apply it self to the Glass, and stick on to the outside thereof like Snow, and tover it all over; and thicken more and more, insomuch that in a short time it will have a thick and hoary beard, all about consisting of natural Ice; and will so long keep on its encreasing as the concentrated cold lasts in the Glass. Then at length the Glass growing a little hot, after the internal cold of the concentrated Fire is consumed, that Icy beard begins again to melt and being resolved, to distill into a Water, for the receiving of which distilling drops some Vessel is to be set under. This is a wonderful Distillation of Men's breath, which coming out of their mouths in their discoursing, is reduced by the concentrated Fire of Salt into Ice, and at length, again into Water by the heat of the Stove.

This so speedy an operation or transmutation of a moist and watery vapour into natural Ice, seems indeed at the outside view to be but a vile and unprofitable thing; but if it be but well minded by the sight of the internal mind, it not only begets a most great admiration, but withall opens the most excellent knowledge of natural things.

Such as greedily hunger after Gold will say, what benefit comes from these tricks? Had Gold but distilled from the Glass we would have saved it, what need we any Water? Or if it had been noble or generous Wine, we could have prized such an ingenious knack, and drunk it off. Take away that filthy Water and bring us the gallant Wine. Such discourse as this, let one of thy Friends purposely utter, being thereto first suborned by thee, that so thou maist the more delight the rest of thy Friends that are ignorant of these things, by thy presently satisfying him that is so desirous of Wine, saying, that if thy Friends and Guests do desire better Wine, thou art ready to draw it them. Upon this, thy Guests will diligently listen and desire to see what better Wine thou wilt draw them out of thy Cellar. The chiefest of these will well know that thou hast not in thy Cellar such variety of Wines.

In the mean while, have ready some small Glasses which contain some Ounces, filled with the concentrated Fires of Salts, and well shut and strings tied ready unto them; now when thou hast a mind to give them a relish of thy Art of bettering Wines, and rendring them more generous, command a Cairn of common Wine to be brought thee, and give it to thy Guests to drink. But now when they shall perceive that it is the same sort of Wine they had formerly, and that thou hast not given them any better, thou shalt satisfie them by the following way.

XLII. A momentary operation of rendering any common Wine more generous, and exceedingly bettered by the cold Fires of Salts; and that in the presence of many Men, Command one of those Glasses prepared for this purpose to be brought unto thee, and let it down by the thread into the Glass full of Wine, which being done, the concentrated cold that lies in thy little Glass, which thou hangest in the greater one of Wine will draw to it self the watery and unprofitable parts of the Wine, and change it into an incipid Ice.

And by how much the longer you leave that little Glass in your Wine, so much the more Water will be drawn there out of, and the Wine will be made the more generous thereby. But the sooner you take it out, the less Water will be separated; so that out of one Cann of Wine you may by this means give your Guests several sorts of Wine to drink, or rather may let them better the Wine themselves even according to their pleasure. For by this operation the unprofitable Water being drawn out of the Wine and turned into Ice, is separated and taken away; part therefore of the Water being taken away, the remainder nn.tst necessarily be much more efficacious and more sweet than it was afore, when it had Water conjoined as yet with it.

A Master of a Family using this Artifice may make for himself and his Guests, divers Wines though drawn out of one Barrel.

Now such a secret is not only full of Curiosity, but also of profit, and may prove helpfull and do much good several ways. I could if need required declare a thousand conveniencies, and Commodities proceeding therefrom, But because I judge it needless to spend time in declaring them, I will at present mention only some few, remitting the rest to the following Centuries, in which shall be made mention of them according as the (matter and) time requires or permits.

XLIII, The amending of any midling or smallish Ale in the Winter Season; as well at Home as Abroad.

It sometimes happens that a Master of a Family hath but only one sort of Wine or of Ale in his Cellar, the which he is accustomed to drink, and puts not in his Cellar any better Wine or Ale, either by reason of poverty, or else because the Cellar lies open to every body, both Men servants and Maid—servants, and they will to the best Tap, and so he fears it will be too chargeable.

But forasmuch as old Men's stomachs, when they sometimes feed on Stock fish dried, or on Marrelmas Beef, or Fish, by reason of its debility through old Age, cannot perform its office of Concoction: The Ale or Wine may be the help of this secret be presently rendered stronger, especially in the Winter Season, in which Season a warmer and stronger draught of Ale and Wine is more beneficial than in the former months, and then they can better brook the want of the same. But some may object and say, where shall I get such a concentrated cold as may enable me to extract the Water out of the Wine? Hereunto I answer that there will be many that will prepare it for time to come and will spare it to others; and yet no body needs so great a quantity thereof neither. If a Master of a Family hath but one only half pound of the same, he may use it his whole life time, if he but keep it so as that the Glass break not and spill it. For when he hath taken away the Water of one or two Pots of Ale or Wine, let him remove the Ice from the Glass, and set again in the cold till he needs it. For such a cold concentrating Magnet always keeps its virtues, and is never corrupted, but always fit for the effecting of many wonderfull things.

N. B. If you have not those fires of Salts the heavy Oil of Vitriol, Oil of Salt, or AQUA FORTIS may be used hereabout; but yet these Oils do not in any comparision perform what those concentrated Fires of Salts are able to effect. But however they demonstrate the thing it self though they bring no great store of profit, and this any one may easily understand.

For there is a great difference betwixt the watery and not watery Fires of Salts, any common and simply bare Water cannot become so cold as the Water of any Salt, and this Salt—water cannot be so cold as a common Spirit of Salt, nor can this Spirit by any means arrive to that degree of cold as a concentrated Spirit usually attains to. So a skin of Leather is never so cold as Wood, nor Wood as a Stone, nor a Stone as an heavy Metal; the diffecence proceeding from the thickness of the compaction, for verily anything will concentrate the more cold or heat and fix it with it self, by how much the compacter and thicker body it shall be of. For it is the property of a concentrated cold to kill a thing and to make it hard and stiff. Contrarywise a concentrated heat gives a speedy life, and correction, and emendation, and this experience it self teacheth.

O happy Man is he that can make a Metalline Salt as compact and thick, and heavy as a Metal, and can by conserving it a due time in the heat of the Fire, that the heat may by little and little and gradually be concentrated and fixed therein, make it fusile. Without doubt such an one would get a Tincture that would cure the most grievous Diseases, and change the imperfect Metals into perfect. For it is the Fire only that begets a maturity to any thing, and by how much the stronger and greater the Fire is so much the speedier and better amending of any thing may be expected.

These things which I have here briefly declared are of greater moment, dignity, and weight than any one can believe; and besides there's no doubt but that there will shortly some step forth, who will without any fear testifie the verity of Art, by changing imperfect Metals and turning them into pure Gold; so common will Alchemy become in this Age, which was neither heard of nor seen before in this World. Nay more, Men will make this Art so familiar unto them that they will not much esteem even of particular Tinctures.

But why God permits such things to be done, is to us wholly unknown, thus much we see only, that doubtless there will follow some great change in the World; happy shall they be who having the fear of God before their eyes, and are of a pure mind, cannot be hurt by the

Devil nor Sin his Mother.

XLIV. Wherein this secret is beneficial to those that travel in the Winter Season.

Necessity doth sometimes enforce old Men to undertake a Journey in the Winter, which, if no urgent hast forceth, may be so ordered that at Noon and Night quiet rest may be always taken in, such a place in which is plenty of Meat and Drink.

But if so be that an urgency of occasion requires a going on forward, whether one ride on Horseback, or in a Coach or Wagon, and that either the Snow render the way difficult, or some Wheel of the Wagon be by chance broken, and so the journeying Person hindred from coming to the place aimed at in the appointed time, he is sometimes by this means constrained to turn aside to a poor Peasants lodging, or if his fortune be a little more favourable, he is nessitated to Inn in some poor Village, where he can neither meet with Wine or good Ale; he now that thus journieth may out of the poor Wine or Ale make himself better Wine or Ale, and the better provide for his health if he hath about him, such a magnet in some small Glass that attracteth Ice to it.

XLV. What profit those that sail in the Sea may have by this secret.

It may so happen that a Man taking Ship with hopes of arrMng in a short space of time to the end of his Voyage, though he has some little of good Wine or Ale, may be enforced if the Wind prove contrary to stay

208

longer upon the Sea; his good Wine therefore and his Ale being spent, he may make that small Beer in the Ship which the common Marriners drink of, better, and preserve his own health.

XLVI. How by the help of this secret the unprofitable Phlegm of Brandy made of Corn may be taken away, that so it may become equal to the Spirit that is made of the lees of Wine.

To the effecting of this business there is required a greater Magnet, which may remove that Phlegm then needed to the Wine or Ale, because Brandy is of an hotter nature than Wine or Ale, which do more willingly let go their wateriness than adust Wine is wont to do.

XLVII. By what means the superfluous waterishness is to be taken away from the weaker waterisher Vinegar, that so it may be made stronger.

The waterishness of the weaker or more aqueous sort of Vinegar doth suffer it self to be more easily extracted by the help of that Ice-attracting Magnet, and the rather because it, viz, the Vinegar put on an Icy form much sooner than any other Drinks.

XLVIII. It may be quaeried whether or no this bettering of Wine, Ale, Vinegar, Brandy, and other Drinks, and rendring them stronger and sweeter, may be done in great plenty, or whether it is to be accounted of as a curiosity only?

For answer, verily it is a most excellent secret

most aptly satisfying the curious inquiry of mortal Men, which the World as yet never knew, and yet it can effect such unheard of things, which it is not necessary that they should be divulged.

As touching the plentifull separation of Water from Wine, Ale, or other Drinks (in great quantity) the same may be done and that with profit, and in some places bring no small gain to him who knows how rightly and artificially to accomplish the same. I have done enough as to my affairs in laying it open; we must not boil meat for the slothfull and thrust it into their mouths. Let them get it themselves if they will, and rightly take care of their own matters.

XLIX. Whether or no likewise a great quantity of cold Fires out of Salts may be easily prepared.

For answer, yes, so great a quantity of them may be prepared as a Man would wish for, or as his necessity shall enforce him to desire. But because such cold Fires of Salts are the effecters of admirable and incredible things which the World never knew of, therefore the copious preparing and getting of them deserves to be concealed. Let therefore every one be content with those things which I have published in the first Century; haply in process of time more may follow.

L. How my SAL MIRABILIS can free watery Oils of their superfluous humidity.

Mix one pound of this my SAL MIRABILIS reduced by warmth into a fine Powder, with ten or twenty pounds of good Oil Olive, or new Linseed Oil; the SAL MIRABILIS is to be commixt warm with the Oil, and being well stirred about with it, draws to it self all the Water, and settles to the bottom of the Vessel, from which the clear Oil is to be separated by pouring it off; and all the Water and impurity of the Oil is to be severed from the SAL MIRABILIS, that so it may be recovered and be again profitable for such like operations.

LI. The way of taking off the mustiness or stink from a Vessel corrupted or grown musty by lying, that it may be again fit to put more Wine into.

Smear over the inside hollowness of such a Vessel with the concentrated moist Fire of Salt, that it may be every where wetted, and sprinkle thereupon so much of the SAL MIRABILIS as will stick thereunto. For so that cold Fire of concentrated Salt, with the attracted SAL MIRABILIS will become hard and not run, and stick on to the Vessel; and that said Fire will in a few days space burn up all the mouldiness and stinch, just as if the common Fire of Wood had been used thereabouts. The Vessel being washed with boiling Water is again rendred fit and convenient to put Wine into.

This operation is not here taught for some stinking Vessels sake that is not worth the while, but to this end, that other secrets of greater moment, and which are profitable, may be learned thereby and known. For under these operations lie hidden many wonderfull things, and such as the greatest part of the Readers will not consider. But to what end is it to light up a

Torch before such Men, that are left by God in blindness and darkness, and hath not vouchsafed to bestow on them any Eyes.

LII. The manner of preserving all kinds of Fruits, Eggs, Onions, and other moist Fruits of the ground a long time from corrupting.

The sweet or dulcified SAL MIRABILIS is to be well dried by the help of the Fire, and being put in some Vessel with Fruits, Eggs or such like, with a thick and close laying (or bed of one upon the other) doth by its driness so preserve all things, and by its attracting virtue of all corrupting humidity, that for a long time they feel not the least corruption.

LII. Question. Why doth the SAL MIRABILIS, which Corn has been macerated withall afore its sowing, and some whereof is mixed with the Earth, (or sown) attract the Rain, coagulate it, and hold it with it self longer than other Salts?

For answer, this is to be imputed to its most great driness which it abounds withall.

LIV. The preparation of the SAL MIRABILIS, so as that it may become an universal Medicine for all Vegetables.

The SAL MIRABILIS as it is of it self, is by reason of its corroding virtues which it as yet retains plainly unfit for the multiplication of the Vegetables, for that being so used would prove more hurtfull than profitable. Upon this account it is necessary that to

212

one part of it be added two parts by weight, of the best CALX-VIVE, which being moistened with Water and made up into Balls, are to be well heated red hot for an hour, that so all the corrosity being introverted the SAL MIRABILIS may be Alkalizated, and used to the VEGETABLES for an universal Medicine; for it conserves its attracting force, and loseth it not in the heating red hot.

LV. Whats the reason that Wood lying in the Water wherein SAL MIRABILIS is dissolved, is turned into a hard Stone?

For answer, this operation is to be ascribed to the incredible astringent property and nature, that the SAL MIRABILIS is endued withall.

LVI. To reduce an half dead Tree to life again by the help of SAL MIRABILIS, that it may revive and begin again to sprout out.

Mix with the digged up Earth, with which the Roots of the Trees are covered, one, two, three or more pounds of the SAL MIRABILIS, according to the bigness or littleness of the Tree, and again, cover over the Roots with the same, and pour upon the Earth it self, some Rain water, that being thereby moistened, the Roots may the better partake of the Salt that is mixed with it.

By this means, the Tree will attract to it self the Medicine or good nutriment out of the Salt, and will be cherished and refreshed just as a piece of bread or other food being given to an hunger-starved Man restores him his strength again.

LVIII. How by the help of SAL MIRABILIS most hard and insoluble subjects may be very easily dissolved.

Let the nature and property of a Charcoal of Wood be considered, the which is such, as that if it be kept in the greatest Fire for many years, and all external air kept out from it; it will neither ever melt, nor ever lose ought of its body, but will come out again in the very same form as it had at your putting it into the Fire.

So likewise a Wood coal is able to endure an hundred, yea a thousand years in the Earth, Water, or even the most sharp corrosive Waters unhurt. This so most sharp a tryal, neither Gold nor Silver though they be the purest and most constant are able to undergo. And although a Coal be thus durable, yet nevertheless will I dissolve it in half an hours space and convert it into a red fusile Salt, which is dissolvable with Water, and yields a wonderfull liquor which is the effecter of incredible operations both in Medicine and in Alchemy.

LIX. What SAL MIRABILIS is to be used to dissolve the Coles.

The SAL MIRABILIS is diversly prepared, as appears in the second part of MIRCULAM MUNDI; but what way soever it be prepared by, it may be commodiously applied to the solution of Charcoals, nor needeth it any further preparation, but even just as it is taken out of the Cucurbit and is as yet corrosive is to be used to dissolve all things.

LX. The manner of reducing any Charcoal in half an hours
space to its first matter, that is, into a sulphureous
Salt, by the SAL MIRABILIS.

Melt two or three ounces of SAL MIRABILIS in some
Pot or Crucible, and throw in a piece of Wood coal or
Charcoal,and cover the Pot with its Cover, and let it
flow for one half hour, that so the Salt may dissolve as
much of that Coal as it can, and may leave the rest of
it which it cannot dissolve, undissolved. Then pour out
your matter and you shall find a red Stone of Salt,
which being tasted upon the Tongue burns it like a Fire,
as all Alkaly Salts do. For the corrosive force is
inverted by the Vegetable Sulphur, and changed into an
Alkaly.

This red Carbuncle being dissolved in Water yields
a green Solution; which being filtred, and let stand
still for some hours, appears of a white colour, and
being let alone quiet longer, acquireth a yellow colour.
One drop thereof gilds over an imperial as Sulphur does,
if it be therein put. For the Charcoal is no other thing
but a Sulphur of the same nature as the Mineral Sulphur
is of, and penetrating all the Metals, suffers it self
to be fixed with them, and doth after another manner
perform all those things that the Mineral Sulphur is
wont to do.

The very well skilled SENDIVOW. in his Dialogue
concerning the Sulphur of the Wise Men, saith he is
strongly guarded, and sits Captive in a dark Prison, and

is not easily freed; but Salt gives him a deadly wound.

A Sulphur therefore sits in this black Coal in a dark and obscure Prison, shut up with strong Bands, and is a Captive, nor can any one free him from those Bands but only Salt. But being once released out of Prison, he is wont to come in view, and not before.

Thus now have we brought forth Sulphur out of his obscure Body. And now will we also bring him forth to publick view.

LXI. How the Vegetable Sulphur is to be made visible.

If you pour into the white Solution of the Coals some Acidity, as Vinegar, Spirit of Salt, of Vitriol, or some AQUA FORTIS, and that leisurely and by little and little as much as is requisite for the killing of the SAL ALKALI; the Sulphur will settle to the bottom like a white Powder, which being separated from the Salts, and washt with fair Water, and dried, will burn and exactly answer to the virtues of the Mineral Sulphur.

LXII. Another way demonstrating that a Mineral Sulphur lies hidden in all Vegetables.

Put this green or white juice of the Wood or Coals expressed or squeezed out by the Salt, in a Glass Cucurbit upon some SAL ARMONIACK powdered, put on an Alembick and draw off all the moisture by Distillation, in which Distillation the spirit of the SAL ARMONIACK will bring over the Helm, the Vegetable Sulphur of a golden colour. It is a most penetrative Spirit and of wonderfull efficacy in Alchemy and Medicine, and this will easily be credited by him who knows its penetrating and graduating virtue and property, in which it excells all other penetrative Spirits, you must keep it very warily because it easily vanisheth.

LXIII. There is yet another way of making the same Sulphur of Coals visible.

When you have poured out our Carbuncle out of your melting Pot, beat it into Powder and mix therewith half its weight of SAL ARMONIACK powdered, draw off by a Glass Retort, both matters exactly cominixt by Distillation, that the SAL ARNONIACK may bring over with it that Sulphur. Wash off this red matter drawn out by Sublimation, with common Water, the which being freed from the SAL ARMONIACK, is a Sulphur inclining from its reddishness to a yellow colour, and is altogether like to the Mineral Sulphur.

LXIV. There is likewise another way of extracting the same Sulphur out of Coals.

First of all, melt the Coals by the SAL MIRABILIS in a melting Pot, that the Salt may be accurately Alkalizated by the Coals, and burn the Tongue like Fire. Then pour it forth and beat the Coals into Powder, and put them in a Glass, and pour upon them Spirit of Wine freed from all its Phlegm. Then set the Glass in warm Sand and ever and anon take it out and shake it well that the Spirit of Wine may extract the Sulphur, and leave the Salt untoucht. Your Spirit being as red as blood, pour it out into another Glass, and again, pour

217

on more Spirit of Wine upon the matter, and repeat the former operation; these pourings on, and catchings off are to be so often repeated, till the Spirit of Wine when poured on will extract no more. put all these red extractions into a Glass Cucurbit, and separate the Spirit of Wine by a B. and it will leave behind in the Cucurbit a sweet Oil of the colour of blood; a Medicine of so great moment in all Chronical Diseases, as that none is to be preferred afore it. For this Sulphur is far better than the Mineral Sulphur, which for the most part is mixt with some Arsenical property, whereas this is extracted out of the Coals of Wood, and is therefore far purer and necessarily more conducive to health.

And as touching the whole operation of this precious balsamical Sulphur, which is but little inferiour to potable gold, the chief knack of duely making it consisteth in this, viz, that the SAL MIRABILIS be well and accurately Alkalizated by the Coals. For if not, the Spirit of Wine would dissolve the SAL MIRABILIS, and would not extract the Sulphur, nor would it answer thy wishes, if it be not deprived or dispoiled of all its humidity.

He that shall be well skilled in the due handling of this Operation, will obtain a most excellent Medicine not much inferiour to potable Gold, of a sweet and pleasant Taste, and of an admirably gratefull Odour and Colour. By such a means as this, is extracted out of a dead Herb, or dead Wood, its greenness in the first Sol- ution made by Water; and after the Extraction with Spirit of Wine, the most delicate red Colour thereof, with a most sweet vegetable Odour; all which lay hidden in the black Coal, and are again brought forth to light.

The use of this most delicate Oil of Sulphur is not small both as to the metalline Operations and other Arts; and this so speedy a putrefaction and revivification of the dead Vegetables into a living medicine carries in its Intrails a great mystery.

LXV. It may be demanded, whether or no the Coles themselves are to be only made use of for this revivification of the dead Vegetables, and not the green or dry Wood they are made of, and the Herbs too, may also be thus dealt with.

For answer, even the Herb it self, or the unburnt Wood it self may be changed in a Crucible into a red Stone by the SAL MIRABILIS. For the operation tends to the same end be it Herb or Wood, green or dry, or made into a Cole.

LXVI. It may be queried, what Wood or what Herb being changed after that same manner by the SAL MIRABILIS, yields the most excellent Medicine.

For answer, the Woods that are weighiest do excell all others, for they are riper and have in them a better Sulphur, than those Woods or those Herbs have which are lighter, and grow up in half a years time, the older the Trees are, the more fit for medicine they are; such as are the Roots of Vines, Juniper, Box, Beech, Oak, Cedar, and such like.

LXVII, A Demonstration, that out of dead Herbs and such as are again restored to life, may new Herbs be produced without the addition of the Seed of other Herbs.

Fill some Pots with some Fertile Earth or Clay, void of all Herbs or Seeds, and moisten it with the green or white Juice of the Coals, If now you expose these to the Sun and Rain, there will spring up thencefrom divers new and unknown Herbs.

LXVIII. How by the help of SAL MIRABILIS, Metals are to be dissolved by the dry way, and to be converted into most excellent Medicaments, and first of Gold.

When you would make your trial of Gold take a piece of golden Money, and bow it, and add thereunto so much SAL MIRABILIS as may be 5, 6 or 8 times the weight of the Gold. Melt it in a Wind Furnance, and pour it out into a Vessel fit for to receive molten Metals; and you shall find your Salt to be of purple Colour. If all the Gold should not be dissolved but some part thereof should settle to the bottom, separate that REGULUS from the purple Salt, and dissolve your remaining Gold in a Crucible with new SAL MIRABILIS, that so all the Gold being dissolved may colour the Salt with a purple Colour. With this purple Salt may be performed many very profitable things, which appertain not to this place. 'Mongst which those are chiefest which respect the emendation or bettering of Metals, concerning which, I will here add only one Operation.

LXIX. The graduating of any Iron into Gold by this

purple Salt.

For the due performing of this, you are to have stone-like melting Pots, and the best that can be, such as by no means may drink in the Salt, or let it run through, for that the Solution of the Gold with the Salt is otherwise wont to hide it self in such Pots as are not strong enough.

If thou canst not get such It is better for thee to abstain from this labour, than to lose thy Gold, unless haply thou hast a mind to try the possibility of the same.

If therefore thou desirest to encrease the QUINTA, (or small weight) of thy Gold which thou hast added to thy Salt, with some Augmentation; put two or three QUINTA'S of Iron bits or pieces into a good Crucible, and having put thereto your purple Salt, melt it very accurately for one half hour, in which time, the Gold will precipitate it self out of the Salt into the Iron, and graduate some of it by turning it into Gold. For whilst the purple Salt doth eat upon the Iron and consume it, it doth together therewithall make some of it participant of a golden Nature by graduation.

I do not insert this Operation here, to the end that by the help thereof a Man should think of getting Masses of Gold, no; for the sole end of my proposing it was this, that I might confirm the possibility of the thing by ocular Demonstration.

Now as here the Iron is graduated by the help of the Gold or golden Ferment, into Gold; so likewise may Copper be graduated and exalted into pure Silver; by the application of a silvery Ferment, as followeth,

LXX. The manner of exalting Copper into Silver.

Dissolve Silver in a Crucible by the SAL MIRABILIS, made of Salt Peter and Oil of Vitriol; in which Solution you shall get a green Salt, fit for the graduating of Copper into Silver, after the same manner as we taught but now of the Gold.

And albeit the Silver augmentation be not so great, yet the possibility of the Art is thence apparent and demonstrateth, that one Metal admits of being converted into another. But yet he that has good skill in handling this labour, will, if he be fraught with good and apt Crucibles, which can hold the SAL MIRABILIS and not swallow it up, receive no small benefit by this same Operation. The SCORIA which are remaining in this, and the precedent Operation are not to be thrown away, but to be mixed with Litharge, that so being reduced by blast, they may graduate the Lead, and enrich it with no contemptible Portion of Gold and Silver. For great are the Virtues of this Salt in graduations, which the Ancient Philosophers have openly enough hinted at, saying that their Salt augments the redness of the Gold and whiteness of the Silver, and that this is a thing most true, he who shall in a due manner perform the Operation will learn that so it is, by his own Experience.

But least an Errour should be committed and some of your Gold and Silver lost, it is better that a Man exercise himself in making his Experiments in the lesser

Metals; and omit the dealing with Gold and Silver so long till these lesser Metals make him a sufficinetly experienced Master for the dealing with the greater.

LXXI. How Iron may be exalted into Copper in the melting by the help of SAL MIRABILIS.

Dissolve one or two whole Lots of Copper in melting it by SAL MIRABILIS, which Solution will yield thee s Salt enclining from a green, to a black colour.

Into the same Pot which contains your Copper dissolved by the SAL MIRABILIS, put three or four Lots of bits of Iron, and adjoin them to the Copper dissolved in the SAL MIRABILIS, and force it with the Fire, so that they may be kept in flux together for one half hours space. By this means the dissolved Copper will adhear to the Iron by precipitation and exalt some particles of the Iron into Copper. All being well molten, pour it out into your Cone that the Copper may settle in a REGULUS. The SAL MIRABILIS and Iron being turned into a SCORIA, are usefull for the inriching of Litharge, in the strong melting by blast, with Gold and Silver.

N. B. If the Mercury of Saturn be mixed with these, or else with those other SCORIA'S which were left by the Gold and Silver and are far better, and so be melted together with a strong Fire, the Lead will be bettered and that by an encrease not to be contemned, and will abundantly pay for the labour and costs. But yet I would not put any one upon the undertaking of this work; except he be well versed with meltings in

Crucibles and without them, by the Bellows upon Hearths. For I write not these things for young Beginners, but only for such who well know what belongs to the Art of Melting.

But yet that he may have some manuduction into these labours, I will declare the general use of the SAL MIRABILIS in the emendation of Metals.

LXXII. The universal use of SAL MIRABILIS in the emendation of Metals.

Though the wonderful Salt of Art dissolveth all Metals, and conjoineth them in a spiritual manner as it were, and renders them efficacious to better each the other in the Fire, yet is there a difference to be observed in that thing, by him who desires to follow the nearest way, and to decline all diversions, or goings about. For example.

He that has a mind to dissolve and conjoin the Metals, Gold, Silver, Lead, Copper, Tin and Iron, by the SAL MIRABILIS, that they may display their virtues in operating to the perfection of each other must take for the Gold, Iron Copper, and Tin, such a SAL MIRABILIS as being prepared of common Salt doth easily dissolve those Metals. But now the same Salt used about Silver and Lead, would effect nothing as to their Solution, because there is no familairity or friendship between common Salt, and Lune, and Lead, for it is an enemy to those Metals, kills them, and reduceth them to nothing.

N. B. But when those Metals are by the help of Salt reduced into their Mercuries, then may it come so to pass, as that they may be conjoined with Gold, Iron, Copper, and Tin; for without a foregoing preparation, they enter not into the Salt, unless the LUNE and SATURN be dissolved in that SAL MIRABILIS which is prepared of Niter, and be adjoined to the Solution of MARS and VENUS; of which Solutions the one doth very willingly embrace the other, insomuch that one Metal doth easily operate upon the other, and consequently a profitable graduation, fixation, and emendation succeeds,

But now if you would have your Metals, not constant in the Fire, but volatile and made flying, then, that SAL MIRABILIS is to be used, which is made of Salt Peter or Kitchin Salt, by the help of Sulphur.

This is the universal use of SAL MIRABILIS, serving for the Solution, Graduation, Fixation, and contraiwise Volatilisation, or the conversion of all Metals into a volatile nature. But the special Solutions, Fixations, or Graduations of them by the Salt of Art, require an addition of some Vegetable Sulphur, which being adjoined to the Metal, yields some help to the SAL ARTIS in the Graduation of a baser Metal, which help the conjunctions of Metals that be of a sulphureous nature, do not at all need; though indeed 'tis better if you help them with some Vegetable Sulphur, For Sulpher and SAL ARTIS are like Male and Female, they bear a mutual love to each other, and beget a rich Off-spring; when they lovingly court each other in the Fire, even alone and without the addition of any Metal, and are brought unto perfection. So then, thus by these operations may gain and profit be divers ways, and in divers manners gotten, as well particularly, as haply also universally, (but this last way I am not as yet acquainted with).

All these things do sufficiently, yea abundantly shew one the way of arriving by the help of the Salt of Art, to the attainment os such things as are of some moment. Enough to the wise.

Although that the Metals do admit of a most easie Solution by the SAL MIRABILIS in the dry way, yet notwithstanding, that Solution may yet more commodiously be perfected the following way, viz, thus, when the Metals are put into the Cucurbit or Retort, in the distilling off the Spirit. For whilst the Spirit is driven off, out of those distilling Vessels, the Metal is dissolved during the Distillation, and remains in the bottom with the SAL MIRABILIS. But whatever of the Metal remains undissolved, is to be removed; but the golden Lunar, Venerial Salt, & etc. is to be kept for such uses as it is necessary for.

N.B. If so be any be minded to pour on again that distilled Spirit, upon the Metalline Salt abiding in the bottom he may so do, and 'tis profitable; because that Metalline Salt is by this means rendred far more commodious and apter for Transmutation.

But that none may err from the right way, it is necessary that we first shew how the Metals (out of which being bettered, Gold and Silver are to be extracted) are to be afore prepared, that so they may admit of being the more easily exalted and amended. For all things are to be done by the prescribed method, and to be managed by (promoting them to) their appointed limit and scope, if any profit is thence sought.

LUNE and SATURN do not by any cominixion associate themselves to the Metalline Salt, unless these same metals be first reduced into their Mercuries, concerning which thing we have treated at large in the third and fourth part of the Prosperity of GERMANY.

LXXIII. By what means the imperfect Metals may (by the SAL MIRABILIS) be turned into perfect ones.

Take of SAL MIRABILIS four ounces, the filings of VENUS half an ounce. Put this matter in a strong double, and well covered Hasslack Pot, set it in a wind Furnace, melt it strongly for half an hour, that the SAL MIRABILIS may rightly dissolve the Copper and make it spiritual. To this Copper thus made spiritual, add half a part in weight of the Mercury of Saturn, and melt together both metals by a repeated melting, for an half or even an whole hour. In this conjunction and operation, the spiritual Copper will get to the Saturn by graduation, no small bettering and fixation. For by how much the longer they are kept in flux, so much the greater amendment doth Saturn purchase. But yet no REGULUS can thus per se settle to the bottom, unless some Iron be added in the melting, concerning which, no certain weight can be prescribed. For when some small bits only thereof, or some little particles are put in the Crucible, the Salt is mortified by corroding of the Iron, and lets fall the amended Lead which in the Cupel leaves the Gold and Silver.

This way shews you the manner of using the SAL MIRABILIS for the amending of metals,

N. B. Other metals may also be rendered spiritual by the SAL MIRABILIS, by which not only SATURN but likewise LUNA may be graduated, or exalted to a golden degree. But yet SATURN is more commodious and fitter for this operation than LUNA is. For when the MERCURY OF SATURN is graduated and reduced, there needs no other labour than the separating of that REGULUS on the Test, where the Gold and Silver is left behind in the Cupel. But if the LUNAR MERCURY be amended by graduation, and precipitated into a REGULUS, it is first to be separated by Lead in the Cupel, and afterwards the Gold and Silver are to be separated by AQUA FORTIS; so that there is requisite a twofold labour, which in the operation by SATURN is but one; and therefore it is to be preferred as to these operations before LUNE.

N. B. All such subjects as have a graduating virtue, as LAPIS CALAMINARIS, HEMATITUS, SMIRIS, GRANATE, TALK & etc. may be used to these labours. But however Gold is the best of all, which if so be that any one is minded to use, it behoves him to be furnished with such Pots and Crucibles, which do not drink up the Gold, and so rob you of more than the produced gain amounts to.

Thus have we demonstrated the use of the simple SAL MIRABILIS, in bettering of the metals.

If some Sulphur be added to the SAL MIRABILIS, it exalts the metals with a far more profitable graduation, and brings more gain than that single operation, which is instituted by the SAL MIRABILIS simply and alone per se.

LXXIV. The manner of conjoining Gold contrary to its nature, with any burning and Volatile Vegetable Sulphur, and of amending the other Metals, all done by the help of my SAL MIRABILIS.

Every body full well knows that there is no affinity or familiarity betwixt burning Sulphur and Gold, which is a fixt Sulphur; forasmuch as they are exceedingly inimicitious to each other, and yet this enmity may be at length changed into the greatest amity.

For 'tis usual with all such as are wont to separate molten Gold from Silver by precipitation, to use common Sulphur about that precipitation, which by its innate Antipathy thrusts out the Gold from the Silver; the same thing is likewise done in the moist way, when the same Gold is precipitated out of AQUA REGIS or Spirit of Salt, by sulphureous Salts, such as are Crude Tartar, Salt of Tartar, Spirit of Urine, and other Alkali Salts.

These are to shew that Gold hateth and shuns Sulphur worse than anything, as being its Capital Enemy; and yet these most bitter enemies doth the SAL MIRABILIS easily reconcile and convert this so great an enmity into sweet friendship. This operation hiding in its Bowels great Mysteries deservedly, and by all right lies hidden to this ungratefull World, if these Mysteries could be excepted which fell into the hands of mine enemies, in my Laboratory, unwittingly to me, who without any regard had to the Writings given me by way of an Oath under their Hands, do now make merchandise of such secrets, and so basely abuse my good Will. Although the manifold use of this great Treasure hath escaped them, upon this account it hath seemed good unto me to reveal unto the whole World, those things that fell within their reach, that so it may be known to all, that such great secrets proceeded from me only, and not from others, not from those mine enemies themselves.

Take one QUINTA of small weight of Gold, more or less, reduce it into thin leaves or plates, and bow them in the fashion of a Cilinder, and add thereunto six, eight or ten parts of SAL MIRABILIS, which matters you must melt in a Crucible with an accurate and strong fusion: When they flow, throw in some pieces of Coals into the Salt and Gold as they are melting in the Pot, that the SAL MIRABILIS may dissolve the Gold and Coals in the melting. Which usually is done in half an hour or thereabouts. The matter being poured out will shew you whether or no you have well operated for all the Gold, as likewise the SAL MIRABILIS and Coals will be dissolved and changed into a red Stone, that bites the Tongue as if it were Fire.

This Fire and red Stone, is the golden Carbuncle of the Ancients, for it shines in the dark like a burning Coal, and produceth such wonderfull effects in Medicine and in Alchemy, which we have no mind at present to reveal. For this Gold being thus conjoined contrary to its nature with Sulphur and Salt, is by that means unlockt, opened and prepared; as that it may by an easie business (or labour) be made spiritual, and that divers ways by divers MENSTRUUMS, either Acid or Urinous, and be distilled over the Helms, and the pure separated from the impure.

And albeit that I here make use of no Circumlocution and speak no plainer, yet have I spoken enough to the Wise, and have shewed them such a way to go in, as that whereby they may without labour, as 'twere arrive to the most happy and wished end, unless God for some singular causes prohibit it.

But that I may not altogether shut the door of Art upon the Sons of Art, I will teach them the making of a most excellent Medicine out of this Carbuncle.

LXXV. The way of making a most excellent Medicine out of the Carbuncle of Gold.

This Carbuncle is to be beaten into Powder, and the best Spirit of Wine is to be poured thereupon, which may extract the Tincture. This tinged Liquor is to be poured off into another Glass, and more fresh Spirit is to be again poured upon the matter, that it may again extract in the heat more Tincture; these Labours you must repeat so often till all the Tincture is extracted, and the Spirit will be no more coloured. The Spirit being drawn off by distillations in a Bath leaves behind a most red Tincture in the bottom, in the form of a Liquor named COS, for here are present, COLOUR, ODOUR, SAVOUR or TASTE; the Colour and Odour from the Gold and Sulphur; the Savour from the Salt. The remainder which is left after the Extraction of the Tincture is not to be thrown away as unprofitable, but to be converted with new SAL MIRABILIS, and Coals made of Vine wood into a red Stone, by fusion, and to be so long extracted till all the Gold be converted with the vegetable Sulphur into a Medicine, For one only labour serves not to extract the whole Gold by the Spirit of Wine; but the oft repeated labours attain to the end proposed.

Thus hast thou friendly Reader a Medicine of great moment and of great efficacy, in which the most pure parts of the Gold and of the Vine are conjoined, nor can they be other than a most profitable Medicament for men and metals.

LXXVI. How by the help of this Medicament, there may be conferred on the Seeds of Vegetables, such an excellent faculty of growth, that they may be as it were seen grow, and may obtain a much nobler Nature, Colours, Savours, and Virtues, than they are wont to get out of the most stinking Dung.

Mix with one part of fat Lome, Clay, or Earth done into Powder, four parts of Sand, that so the fatness of the Earth may be somewhat allayed. With this mixture fill a Pot, such a one as the Garderners are wont to keep their Flowers in; pour thereupon some Rain-water wherein is mixt (or dissolved) a little of that Medica-ment made of the Carbuncle, and plant or sow in that Earth some of those Herbs which abide unhurt by the Winters cold. Set the Pot with the implanted Herbs to the warm Air, but so, as that no Rain come at it, for the Rain may wash away that medicinal nutriment. When the Earth becomes dry, you must pour on more of the Medicine prepared of the Carbuncle, and that so often as need requires. So will the Herbs begin to grow, which if they meet with no at her nutriment besides the Rainwater, they cannot attract any other whereby their faculty of growing may be promoted and encreased. And for as much as the Golden medicament was adjoined to the Rainwater, the Herbs must necessarily draw it to themselves together with the Water, and obtain other properties than if they grew from the stinking Beasts Dung.

N. B. Under your Pot that contains your Herbs is to be put a Dish made of good and firm Earth, or else of some Metal, which may serve to catch the medicinal Water, that flows through the bottom of the upper Pot, or distills thence, and having received it may not drink it up but conserve it. Besides, it would not be amiss if some of that medicinal Water were put in the under Platter, which might always keep the bottom of the upper Pot moist, and so may supply the Herbs with an uncessant nutriment. It would be better also, if the Pot it self were made of some Metal and not of Earth, that so it may not drink in that precious Water, but rather conserve it.

LXXVII. What is to be observed in this Operation, that some good effect may proceed from thence.

In the first place, diligent heed is to be taken, that the Lome or Earth you take, partake not of any salt faculty, nor hath any other corrosive Property, for many such Earths there be which would hinder and spoil the faculty of growing.

Secondly, there must regard be had to the moistening of the Earth, lest the Seed be choked with too much humidity, or in defect of sufficient moisture, dry up and wither.

Thirdly, there must be observed a measure of the Medicament it self, that neither too much, nor too little of the same be commixt with the Rain-water. For an overmuch quantity thereof burns up the Seed, and a more sparing Portion cannot yield nutriment enough to the Herbs.

Farther, some Musk or other things that emit a fragrant Odour may be therewith mixed, which addition is wont to get to the Herbs a most fragrant Odour, If so be a man fears to apply the aforesaid precious Medicament to this Operation, because of the Costs of the same, he may use that Tincture which we taught a little afore, to prepare of Coles only without Gold; and which indeed will perform all those things, (as in reference to the growing faculty) which that Golden medicaznent is wont to perform; this only excepted, viz, that the Herbs will not partake of that golden Property which they, obtain by the Golden medicament.

The things we have here written and published concerning the promoting the faculty of growth in golden Herbs though they seem not of any great moment, yet hide they under their mysteries of great moment, the which many Artists will apprehend, and convert unto their Use.

LXXVIII, How any Wood or any Wood-coal may be so prepared by the SAL MIRABILIS, as to be capable of a long while resisting the Fire,

Dissolve some SAL MIRABILIS in common Water, put some Wood or Wood coal therein, let it lye in it for some days, or so long till it be well glutted with the Liquor and become ponderous. Then take it out and dry it very well at the fire, that all the moisture vanishing away may leave the SAL MIRABILIS in the Wood: Then put it in the said Solution yet again, and take it out and again dry it, which labour will render the Wood so much the solider by how much the oftener it shall be repeated. By this means, all the Pores will be filled with the Salt and the Air will be shut out, that it can penetrate it no more; without which Air no Wood can ever take fire and burn. If now you put such Wood or such Coals with other Wood and Coals in the fire, these (un-imbibed Coals, & etc.) will be consumed by the Fire in a short time and be reduced into Ashes, but those others will remain untoucht, and may be taken out unhurt, though indeed even they too will be burnt if they lye over long ,in the Fire, This is certain concerning Coals, that those that are made of more weighty Wood, and which abound with a greater Quantity of Salt, such as are the Oak, Beech, Juniper, Vines, and other Trees whose Wood is ponderous dure far longer in the Fire, than those Coals do which are made of Firr, Pine, Alder, Willows, and such like lighter Trees, and which have a lesser Quantity of Salt, and this now I do not mention barely for fashionssake, but to this intent, that occasion may be given from this kind of knowledge of drawing some profitable matter therefrom; as for Example.

LXXIX. How such kind of Woods which 'are always so near the Fire, as that they are still in danger of being burnt, and thereby threaten damage may be conserved from firing.

Dissolve some SAL MIRABILIS in Water, and with a Pencil smear over such Wood which by reason of its nearness to the Fire is always in danger of being burnt. When the Water is dryed up, moisten it again with, the same Water, and repeat this moistening so often, till it hath drunk in a sufficient Quantity of the SAL MIRABILIS, and become able to resist the heat. By this means might men be often freed of many fears and cares, in ships dawbed with pitch and in other places, where by reason of the too nearness of dry Timber there is danger of firing.

LXXX. How by the help of SAL MIRABILIS any Wood may be conserved so, as for a long time to remain unhurt in the Water.

He that desires to preserve Wood, that it may not be detrimented by the Water nor rot in a long time, may be master of his wishes, the following way. Dry your wood very well, and being dry moisten it with strong Oil of Vitriol very exactly, and being moistened sprinkle it with the SAL MIRABILIS beaten into Powder, that it may stick well on to the Oil of Vitriol, For the Oil of

Vitriol doth in its penetrating of the wood carry in the Salt thereinto, and makes in the outside thereof every where about a black Crust, just as if that wood had been burnt by the Fire. Now because Coals resist putrefaction, it must necessarily follow that the wood being in that wise ordered must remain a long time unhurt in the water.

LXXXI. The Preparation of the SAL MIRABILIS for this Work,

There ariseth no small difference amongst the Salts themselves from the different way used in making the SAL MIRABILIS.

If the Oil of Vitriol wherewithal this SAL MIRABILIS is prepared, be not by reason of the superfluous humidity strong enough, any one may easily conjecture that a good SAL MIRABILIS cannot be made thereof, because the Kitchin Salt would receive therefrom but little alteration. To prevent this in convenience therefore, you are to take equal weights of Salt and Oil, that so one may be assured that the common Salt is well inverted, and made a good SAL MIRABILIS.

LXXXII. By what means trial may be made, if the SAL MIRABILIS be duely prepared, and how it may be fitted for this and other Uses.

Its Colour ought to be white and transparent; its figure is in long STRIA'S or Chrystals; its taste is like Ice melted upon the Tongue and yields some bitterishness. Being th"yed in the Fire and all the

237

moisture gone off, it will loose three parts of its own Body, and retain a fourth Part only, being dissolved in Water it will recover those three Parts again.

But on the contrary, if it shoot into a square Figure, and hath as yet a saltish taste, and being dryed loseth but little of its weight; it is not worth a rush, and shews that either the Oil of Vitriol it self was not good, or that there was not enough used to the Operation. These things we would not bury in silence, that so we might well advise young beginners, and withdraw them from their Errours.

LXXXIII. It may be quired, whether the SAL MIRABILIS serves for the use of Artificers and Craftsmen.

For answer, yes. For this SAL MIRABILIS is not only able to perform things of great moment, and those too, such as are not common; both in Alchemy and in Medicine; (a rehearsal of which, we Shall for brevities sake omit) but withall it may be used in other Arts and Handycrafts with great admiration and profit; and this we cannot neither at this time demonstrate because of the but now mentioned brevities sake. We will only shew here, that even the poorest Husbandman, might (if they knew its preparation use it to notable advantage and profit.)

LXXXIV. How every Countryman may encrease any kind of Corn or Seed with a thousand fold encrease by the SAL MIRABILIS if he can get it.

We have aforetold you, that the SAL MIRABILIS

being so, as it is per se, is plainly unapt for the multiplication of the Vegetables, unless that corrosive Faculty be taken therefrom by lime or other ALKALI Salts, (the which must be done) if you would expect thereform any good concerning this multiplication. Here now will I disclose a business of no small moment; yet not to this end as If I would perswade the Countrymen, to get for the future, or afford to their Corn so plentifull a faculty of encreasing. No, no, I well know that they know not how to make the SAL MIRABILIS, and if they did, yet would they not depart a Nails breadth from their Ancient Custom. For 'tis a conunon Proverb, Old Dogs are very difficulty tamed; and this, the common Course of mens Lives doth clearly teach; wherein you'll find, that a man hardly unlearns that in his old Age which he learned in his Youth; so that an Old man doth very difficultly suffer himself to be withdrawn from those things whereto he hath been accustomed when young. Neither is it my purpose so to do, for as much as I insert the things here mentioned by me, for this end only, viz, that the possibilty and wonderous Properties of the SAL MIRABILIS may be brought out of Darkness into Light, and may be made evident to the whole World.

I would likewise be thus understood as touching other Workmen, for whom these things are not delivered or treated of, that they should desist from their old Custom, and obey my admonitions and instructions; but for this end it is only that I publish these things, viz, that they every one may know that my SAL MIRABILIS can be an helper to men of all ranks, and also may bring even to the Craftsmen themselves and to the poorest Husbandman, great fruit and benefit.

Now when you hear it mentioned that some Grain of Corn is augmentable beyond the usual Custom, by an unheard of multiplication, it must of necessity be, that it emit more than one, two, or three Stalks, for as much as so few Stalks, cannot yield so great an Encrease, But now if one Grain is to put forth so many Stalks, it is wholly necessary that it be done by some certain singular and strong efficacy of expulsion; and that too, even presently and at the beginning when the grain is at first sown in the earth. For whatsoever is not here done even at the beginning, will never be done afterwards.

For all the stalks that spring forth after are small; and quite unfit to bring forth Corn, So then, seeing that many stalks are to break out at one and the same time out of one grain, if an eminent multiplication is expected to follow, then verily 'tis even necessary that some help be adininistred to that same grain afore it be put in the earth, that so it may plentifully grow and be speedy in presently sending forth even at the very beginning; good store of stalks.

The Countryman know not any thing serving to such an operation but only Dung; but I do even now again say as I have often done afore, that this effect of the usual and common multiplication ariseth not from the Dung it self, (as being but the outside Husk) but from that sulphureous Salt that lies hidden in the Dung. Hence it is, that by how much the purer and better the Salt is so much the speedier and more efficacious an operation ariseth therefrom. If then, that such a Salt can be made by Art which performs the same that Dungs does, it altogether follows that we are able to do the same without Dung, and that far better than by Dung in which the salt is so much dilated and which (by the benefit of Art) we contract into a narrow compass.

I hope the well minded Reader will not be displeased that I use so many words here, about the stinking Dung of Animals; because I can't indeed use in this place any other manner of speaking; seeing I intended the laying open of this thing and there fore am I even compelled to speak of the same: For he that minds the publication of any thing, cannot do it, unless he speak of the same. And although that Dung may seem to some finical Men a very contemptible thing, yet notwithstanding it is the only and principal MEDIUM, by the help whereby our daily Bread and the necessary sustaining of our Bodies is had. But as for the stinking Dung of Animals, I even remit it to the Dunghill, and return to my SAL MIRABILIS.

LXXXV. Whether or no a thousandfold encrease may be had of Corn by the SAL MIRABILIS.

Melt one or two pounds of SAL MIRABILIS in a Crucible; then throw in some Coals and dissolve them, and reduce them by Solution into a red and fiery Stone; which matter being compounded by melting of them both, beat into Powder and pour thereupon common Spirit of Wine, that it may wax red by extracting out the Tincture. Pour this out into another Vessel, and pour on other Spirit upon the aforesaid matter. And this pouring on and canting off, is to be repeated so often till all the redness is extracted. By this extraction you shall get a sulphureous Salt, fit to steep or macerate Corn withall; because it agrees very well to the properties of that Salt which sticks hidden in the Dung of Beasts: Now I use Spirit of Wine to the extraction for this cause, for that it hastens the germination or budding of the Seeds even as well as the Salt doth, and enricheth it with an. emission of many Stalks. But yet your Spirit of Wine must not be over strong for then it would hinder the faculty of growing, the which thing even the Salt will also do if too great a quantity thereof be added to the weaker Spirit of Wine, because it would by burning up the Seed be an impediment to the faculty of the Seeds growth. It is therefore necessary that a good regard be had (in those operations) to a due measure. For an overmuch access of any thing is wont to be no less hurtfull than a defect or clearly wanting of the same.

This is the preparation of the SAL MIRABILIS necessary for the macerating of Corn that so it may produce many Stalks; now follows the true and genuine use thereof in macerating of the same.

LXXXVI. The true and right way of macerating Corn in the SAL MIRABILIS.

There are several kinds of Corn, and of these various and different sorts. Hence is it, that one Seed is longer a macerating than another is; and that because one becomes soft sooner than doth another, or attracts humidities to it self quicker than another; so that regard is to be well had to the difference thereof. Rye and Wheat are encompassed with thin Skins and therefore are the sooner macerated, Oats require a longer time, and so doth Barley which has a yet harder Husk than the Oats hath and therefore requires a longer time for its maceration, But as touching these things, every one may find them out by his own understanding and often experience, because it is impossible to mention all things so clearly and perspiculusly. But this is a general rule, your Corn is to be so long left in steep, until you may easily bite it a pieces; for you must beware of softning it too much, for then it would presently putrifie, and by that its putrefaction corrupt and spoil all the growing faculty, But experience will instruct you far more commodiously and more perfectly herein, than a larger description can.

LXXXVII. The true and right way of sowing your macerated Corn in the Earth.

Any one may easily conjecture that if the Corn macerated by the aforegoing way, be sown in the Fields the usual way and so thick as the Husbandmen are wont to do, it will not succeed because of the overmuch thronging and thickness by which the Corn would hinder each other and so choak themselves, This incommodity therefore is to be prevented, and such macerated Grain to be thinly sown in the Fields that they may have room for the freeness of Air, and so may grow up and not spoil each other by a mutual suffocation.

Nay rather that the more accurate diligence may be

had or used about this sowing, a Man may make him some wooden Instrument, whereby together and at once many grains of Corn may be sown in the earth in good order and at a certain distance; concerning which labour I have purposed to speak more at large in another place. For so no grain will unprofitably perish, and with one Sack of Corn may be sown more Ground than six, eight or ten sacks are wont to do otherwise; my too short time constrains me to break off my discourse concerning these things.

LXXXVIII. By what means the SAL MIRABILIS may bring profit to the Dressers of Vines.

If there could be a good quantity had of SAL MIRABILIS, and that without great costs, 'tis without doubt but the Vines might be made very fruitfull therewithal.

But because they are ignorant of the preparation thereof, who dress Vines; 'tis expedient for them to acquiesce in their Beasts Dung, or make use of that only for the fattening and dunging of their Vines, which is made of common or Kitchin Salt by inverting and Alkalizating it by CALX VIVE. Unless a man has Vines about his House or in his Garden, and would make them fruitfuller than ordinary. For to make tryal thereof in great Vineyards would be too costly.

But yet I will propound another way to the Vine dressers, by the help whereof they may get plenty of Wine every year. I have at large taught in my foregoing Writings, and that by various descriptions, by what means one may be Master of noble and ripe Wines every year, yea even in those times, in which by reason of the coldness and unseasonableness of the Air, and the want of the Solar beams, the Grapes cannot attain their due maturity; and withall, how in those places where they seldom or never grow ripe, (they may be ripened) by concentration by the help of the Fire, or else by fermentation with their own vinous spirit distilled out of the Lees, and added to the said Wine; (having I say already taught this) there's no need of repeating it again. Yet nevertheless if God lengthen my life out so long, I have purposed to write a peculiar Book of the propagating and bettering of Wines.

But that I may ingratiate my self with those who not much caring for poor thin Wines, desire to have some noble Wine in their Cellars, I will here reveal a certain secret which may not only refresh the body and spirit of many thousands of Men, high and low, rich and poor, throughout all GERMANY, yea and all EUROPE too; but also administer them no small profit.

I have taught a little afore, how by the help of the SAL MIRABILIS and of the concentrated and cold Fire of Salts, a Man may make his Guests different Wines out of one Cup, and therewithall refresh them; and I have likewise shewed, that such a bettering of Wine may be exercised in most places, with notable profit.

The truth of which thing, it hath seemed good unto me to demonstrate divers ways, for the sake of my Neighbour.

I have frequently laid open in my Writings some

excellent secrets, and have withall made a discovery of the most great benefit which one may thereby receive.

But because I have not pointed out with my Fingers where and whence such notable profit is to be gotten, the most part could not apprehend or find the same, and have therefore rejected the thing it self being it was not so perspiculusly and clearly laid open, as unprofitable and worth nothing. And now least it thus happen to this secret, if I should not shew and point as it were with my Fingers, the benefit thereof, and if therefore such secrets lying in the dark should not come forth to the profit and use of mankind, which would be a grievous thing and to be lamented, if it should not, it hath seemed good unto me to manifest the utilities of the same.

LXXXIX. By what means notable profit may be gotten by my Water-attracting Magnet.

First of all, it is no small benefit when the overmuch Water is taken away from the poorer sort of Wines, with which GERMANY doth every where almost abound, and the Wines made nobler, stronger, more efficacious, and more durable, and do get a far more acceptable savour.

For the unripe and watery Wines are not of any long lasting, but do in a while lose all their savour and all their strength, and become mouldy, and corrupt with lying, and 'tis not seldom that they grow tenacious, or ropy, thick and muddy, reddish and filthy; all which incommodities doth my Magnet cure in the space of one only hour, by the drawing away the overmuch waterishness.

XC. The second benefit.

If a plentifull Vintage or large encrease of the Wine should be more than you have Barrels to fill, nothing can be more acceptable than that the Wine may be concentrated by the Magnet, (which draws to it self the Water and turns it into Ice) by extracting the over-muchness of the Water, that so by this means the more contracted Wines may be laid up the more commodiously, and may if not very good, be rendered better.

XCI. The third benefit, and which is most acceptable to all Masters of Families.

If so be that the Master of the House had a whole Cellar full of Wine, and every Vessel filled with the like or self same noble Wine, without any difference; then verily the Master and Servant would be of equal degree, nor should the Master have any prerogative above the Servant. But now using the help of the said Magnet, he may have his Wines bettered as he pleaseth, and have divers Wines in his Cellar, as we taught a little above; if, viz, he shall draw from it the unprofitable watery part.

XCII. Another way of getting profit by the Magnets drawing the Water out of Wines.

The Anatomizing and examen of Wine discovers, that in twelve measures thereof; there is about one of more noble spirit, and almost one measure of Tartar, The residue are nought else but an insipid water altogether like to common water. Now when the Wines are to be transported out of the Countries wherein they grow, into more remote places; what need is there of carrying the water with it so long a Journey? Would it not be better to separate some part thereof from the Wine, and so transport the Wine, and let the water alone, and thereby shun a great deal of charges disburst for the carriage of water into such places as have enough already? Would there not rebound a great benefit hereby, both to the Buyers and Sellers of Wines? Yea; verily, I believe that there will not only redound unto them a great conveniency, but withall a great deal of Treasure.

XCII. There's yet another way of getting notable profit by the said Magnet; viz, if the ill taste and fetidness is taken away from the Brandy, usually made of Corn.

There is some mention made in what went afore concerning this amending; but 'tis not done so clearly and manifestly. I will therefore open it more clearly and more perspicuously in this place.

Mix one part of your Brandy made of Corn, with two parts of common water poured thereunto, that the stench and ungratefull savour may diffuse it self into the added water, Having so done; you must again free this Brandy thus tempered with water by putting your Magnet thereinto, and so will you draw therefrom all the stinkingness, and 'tis just as if you had washed that Wine, and rinsed off all its filth.

XCIV. The benefit purchased by separating the Water from the Vinegar.

If you would have benefit by this liquor; the same may be done by the same reason; in those places out of which it is transported into other Countries, if, viz, the unprofitable water be removed after the same manner as we taught to be separated from the Wine,

XCV. By what means good Wine and Vinegar may be every year prepared by the help of this same Magnet, in those Countries which the Grapes do not ripen.

Although that in all the Coasts scituated upon the Rhine; as in RHINGOVIA, MOGUNTIA, WORMATIA, ALGENTORATI; in ATSATIA, the PALATINATE, FRANCONIA, AUSTRIA, and the Dukedom of WARTENBERG, (in which places; the Wines do for the most part arrive to their perfection every year) this Art be not so very necessary; yet nevertheless SAXONIA, MISNIA, THURINGIA, SUEVIA, and BAVARIA do stand in need thereof; in which Countries the Wine doth for the most part remain acid, unless the Summer hath by a singular chance hapned to be very hot. For in these places it is no less profitable than pleasant to drink a sweeter and nobler Wine instead of the more acid Wine, if so be one could get it by the help of the said Art. Besides; this same Art yields no small profit and benefit in those Countries, in which though the Wine is (as we said but now) wont to be noble.

For it may happen that some unseasonable Weather may hinder the ripening of the Grapes, and that the

Vineyards wanting the due heat of the Sun cannot arrive unto maturity. For oftentimes in one and the same Country, there are divers Wines produced; so that one sort exceeds another in nobility and goodness. Those therefore that are good and generous need not the help of this Art, but contrarily the smaller and less noble Wines want it. Hence I conclude that in all the Countries of the World, wheresoever Wines are made, this Art may be serviceable and profitable to any one.

XCVI. How in those cold Countries; as in POLAND, DENMARK, NORWAY; & etc. Which by reason of the Coldness of the Air admit not of making Wine there may nevertheless good Wine and Vinegar conducive to the health of Man be made.

　　Though the Cold may so hinder as that Vines will not grow; nor Vineyards be, yet notwithstanding those places so obnixious to the Cold, have plenty of Apples Pears and such like Tree fruits; whose Juice being pressed out; and fermented; and after the Fermentation, freed of the greatest part of the unprofitable water; will give a better and more durable Drink, than that which is made the usual way of Apples and Pears. For this drink cannot last long because of the muchness of the humidity but becomes ropy acid and muddy, and so corrupts.

　　So likewise may those Countries that abound with Corn be rendred partakers of most excellent and wine like Drink which may be used and drunk instead of Wine, to the great benefit and advancement of the health of the Body; and it is to be thus done.

First of all, let very good Wheat be made to germinate (or sprout) by stewing in some gentle heat; (as in making of Mault) then after the sprouting let it be put in some warm Furnace, or in great Coppers; and stirred about with some wooden thing without ceasing till it be dryed. In this Operation you must have an especial care that the Corn smatch not of the Fire which is then brought by drying to a sufficient hardness; when it is not soft in biting it with your teeth, but leaps as 'twere in pieces; this is a sign that you have done your work well, Having prepared it thus, let it be broken (or ground) and boiled after the manner of other Ale, without Hops; and then after it has fermented let it be freed from its superfluous moisture by our Magnet, So will you have remaining a sweet Drink not much unlike to Wine; which Liquor if you would yet have more near in its likeness to Wine, you must put in for every Pun or every Butt about a pound of Tartar in the Fermentation; that so it may ferment together with the Corn, and may give unto this Drink a winy taste.

After the same manner may excellent Metheglin be made of Honey and Sugar, or such a kind of Drink as but little differs from the Savour of Wine. An Art, verily; most profitable in those places which have no Wine, but have that defect supplied by the greater Quantity of Honey and Sugar. The manner of making such a kind of Drink is this.

Let there be added to the Honey so much water as is sufficient, for their boiling together in a Copper;

and let them being boiled be diligently scummed so long till the Honey becomes thick again, and gets a duskish or a reddish Colour, which is a sign, that the less sweet, and less honied Savour is removed away, by the boiling.

To this Honey reduced to the said thickness; let be again added as much water as is sufficient, that they may be boiled together, and being boiled put up in Barrels, Whilst it is yet warm some Ale yeast must be added; which being fermented, renders this Liquor so sweet that it is but little inferiour to Wine; moreover it will come nearer to the Taste of Wine; if a due part of dissolved Tartar shall be added thereunto in the Fermentation, that so being fermented together herewith, it may acquire to itself a winy Taste.

N. B. But here good heed must be had, that in the last Solution there be not taken too little water, but rather more than is wont to be taken in the making of common Mede. The reason is this; because the Honey gets not its due Fermentation, but retains its usual and a kind of nauceous Sweetness, and cannot be made partaker of a winy Savour.

But now, water enough being added, promotes the Fermentation, and causeth that it gets its Purity and Clarity much sooner, and is of an excellent sweet Savour. After that this Metheglin hath gotten the requisite Clarity, the superfluous water is to be abstracted thencefrom by your Magnet, which water being removed, the remaining ungratefull Savour of the Honey going away together with the water; vanisheth, and this

Wine of the Honey, gets its strong Spirits from the Honey, and has a winy Savour from the Tartar. If any one be so minded, he may add to the Honey in the first boiling; some Spices, or which is better, may hang them in a little Bag in at the Bung, that so they may be fermented with the Metheglin, and give it a sweet savour, The Spices are these that follow: Cardamoms one, Coriander two, Orris Roots three parts; the which will give the Metheglin a fine Taste, The Flowers of Elder make it taste just like Wine made of the Apian or Nuscadel Grapes, Cinamon and Cloves also do give it a delicate Sweetness, But every one may use such Spices as he thinks best, according as he fancieth this or that Taste. Of such Metheglin is made most excellent Vinegar, which scarce is inferiour to wine Vinegar though of the very best Sort.

XCVII, Whether or no, there be any other benefit, which our Magnet can bestow.

He that shall only diligently enquire by trials made, will without doubt, find, that such a Magnet as attrácteth water, can be profitable many ways; which to treat largely of here, the time will not permit, For because that this Magnet draws out of all Liquors, their superfluous water, it doth certainly bring much Benefit, and manifold Fruits, very many of which, we would here declare if need required. But the time admits not of any longer dwelling about these things, But yet, however, I will reveal an Art for the Poor's sake, who have no Wine growing, nor any money to buy it, and are therefore enforced (but especially in the winter Season) to drink cold water after their hard Labours; by the help of which said Art they may have good Wine to drink all the year both in the Summer and Winter months; I mean in those places in which there's plenty of Wine made, and is in the Autumn Season squeezed out with wine Presses.

In all those places in which store of Grapes are prest out with Presses, there is great store of the husks, the which is partly kept for the Beasts to nourish them in the Winter, and partly thrown away as unprofitable, especially in those Seasons which afford a great Quantity of Wine. But if so be that the Wine Harvest be somewhat poor and not so plentifull, then they pour water upon all the Husks or on some part of them, and leave it so for some days, and again press them; and thereby is made a Drink that has some kind of wine-like Savour which is given to the Servants, and other Labourers to drink instead of bare water.

But now in such Years wherein they are thoroughly busied in curing or making much Wine, they have not the time to bestow about making that Drink then Nay sometimes they have such a deal of Wine that they have not Cask to put it in, but are compelled to give away their smaller Wines to others, and stuff their Cellars with the more noble.

If therefore the poorer sort would have now and then a good Draught of Wine, they must get them some large Vessels which they must fill with the Offal of the Grapes, and with water poured thereupon, and leave them thus, so long until the rich People have done with

their-wine Presses and stowed their Wines in their
Cellars. Then may they also press out their second Wines
in, the rich mens wine Presses, and by the oft-spoken of
Magnet separate the unprofitable water thereform, and so
lay up their Wine, the which will last and abide good
and durable all the year about, which it will not otherwise do. For such kind of second Wines dure only (for
the most part) but the Winter and Spring, and part of
the Summer, and the utmost time they remain any thing
good is but till the Month of JULY, afterwards they grow
ropy or musty. But now the unprofitable partof the water
being separated, they get a shorter or longer
durability, according to the moreness or lessness of
their Concentration. This advice and secret was I
willing to bestow upon the poor that they may also drink
good Wine.

But they may demand where should we get us such a
Magnet, by the using of which we might make our second
Wines good? I answer, they may borrow it of the Rich men
who have Vineyards, for so long; and when they have done
with it may restore it them again; for it is so lasting
that it never loseth any thing of its Body nor of its
Virtue, but always remains good. The rich men therefore
will buy it of the Chyinists, the preparation whereof is
clearly and perspicuously delivered in my first Century.
If now this be done (which I doubt not of) the Chymist
by preparing, and the Merchant by using of the same will
reap no small Benefit and profit, And questionless,
there will be found some men that are studious of new
things, who will make trial of this Concentration of

Wines in small experiments, that so they may fish out the possibility of the thing. But I know not whether they will exercise this Operation in a greater Quantity or no. For its neither here nor there to me whether they will exercise this Art or let it alone. For this is the natural disposition of most men, they would very willingly get store of gain, provided it could be done without great labour and much trouble. From hence it is certain, that this Art of Concentrating Wines by cold Fires will not be so soon common, especially because I have not here delivered how such a Magnet may be made in great Quantity, and applied to use.

But yet I think I have sufficiently done (or hinted at) those things I have revealed, Let others draw out of their own store too, if they have any thing. More things I could not reveal, for many weighty causes which I count it needless to mention here. Very many men will commit many Errours, e're they attain the right Scope. Verily it would be much better if there were a greater Number of such Persons as readily understood this Art; especially in those places of GERMANY in which the Wines are so acid, that it will make ones Dyes run with water if a bigger Draught than ordinary be drunk to quench Thirst. And therefore in those places Ale is in the most esteem, and indeed it is no contemptible drink, if good, but yet it is not at all comparable to the noble Wine, concerning which noble Liquor these Verses may rightly be pronounced.

With what a lovely gift are all things blest
By th' noble Wine from tender Vines exprest,
To sick mens pains it doth an easement bring,
It joys the Country Peasant, makes him sing.
And you shall see that that man whom to day
By means of Wine lies tumbling in the way,
Will on the Morrow have his tother lay.

There are many Songs in the praise of Wine, but Ale is not celebrated with any ditty; though it be never so good. Upon this account therefore the noble Juice of Wine is not undeservedly preferred before all the rest; provided it has its due generousness and excelling goodness. But if it be not good, 'tis wholly expedient to help it, lest by keeping its Sourishness and waterishness it perish. But filthy covetousress bears too great a sway upon mortals; insomuch that there are too many to be found that would rather pour water to their Wine and spoil it, than better it by abstracting of the water.

I have often heard the complaints of Vineyard Masters in those Seasons, in which they have had whole Cellars full of small and poor Wines; which have not arrived to so much ripeness as to be able to be sold and transported into other Countries. Hence comes it to pass, that if they lie along while and be not drunk up, they degenerate more and more, and become exceeding poor, and at length corrupt by lying and so perish for altogether; unless it happens by chance that some years of a more happy Vintage do succeed; with which more noble Wines they may mix their small ones, and so sell them off; but yet with a poor profit; because the Merchants are not wont to buy midling Wines, but the very best of all. But they may have noble Wine every year by that means that I have shewn. For if one eighth part of water were extracted from the Wine, it might have the name of good Wine. But now if a fourth part of that water should be taken away thencefrom it would become far more excellent, for a little water is able to make a most noble Wine smaller and more base, a tryal whereof you may make as follows.

A Hogshead of rich Wine, containing some six Renish OMA'S or AULM'S, costs an hundred imperials; yea (sometimes) more, an hundred Duckets: Now if you take thencefrom one eighth part of the measure, and put in the room thereof one eighth part of water, you will find by the taste, that its goodness is so much diminished, that that Vessel will be scarcely valued at fifty imperials. But if a quarter part be taken away, and so much water put in its room, any one may easily conjecture that such a Vessel filled with such Wine will be hardly judged worth twenty imperials.

So then by this way that I have told, any Wine might be brought to such a nobleness by taking away one fourth part only of its water, that a Vessel which afore would have yielded but twelve imperials, will afterwards yield three times the price.

By all this that hath been spoken may any one easily conclude the truth of the true Alchemy, and what incredible benefits may be reaped thencefrom. From this ground I say, that that delicate sup of Wine which I have here taught the preparation of will so inflame many, even of the enemies of Alchemy, that they will for the future put their hands to the Coals, and try to get thereby such a delicate Magnet, and to have it by them.

This Magnet will also help not a few Alchemists themselves that are in straights and wants, and afford them a good Cup of Wine; whose Vineyards, House, and all their substance, the smoak and hot fire hath already driven up the Chimny, and (in lieu thereof) this cold fire will recover them again with no contemptible increase. For this Leap year 1660 is the first year since the World's Creation, in which the miserable Coal-blowers may arrive to a way of getting their Bread, if they will but stretch out their hands. And that I maybe here well understood, I don't only mind the concentration of Wine, for there are other most profitable uses to be found out of these cold—fires; which time will manifest.

It seems good unto me to add by way of an over-plus (because I have taught the making of a Cup of good Wine, and there is a great familiarity 'twixt Wine axid Corn) the showing a way how one may get a most delicate sort of Bread and of sri excellent savour, that so he may be furnished with the choisest of Food and Drink.

Let some part of the water be extracted by the cold Magnet out of new Milk, that it may be made better than the common Milk. This would be a most excellent nutriment for the sustaining of Infants, whose Mothers die too soon. For every Infant cannot bear raw Cows

Milk; and if the superfluity of the Water be removed by boiling, the Milk (not brooking much boiling) doth easily taste of the fire, which would not be if it were freed of its water by the oft mentioned Magnet, for it would remain sweet, and be of an excellent taste, With this fat milk, moisten your Wheat Meal, (which must be of the best) in the stead of water; and let the Bread be baked, and without doubt the Bread will be of an excellent Taste, which could not be by the common baking, although that Butter were added thereunto. One may feed upon this Bread alone, without Cheese and Butter, because the Cheese and the Butter are with the Milk in which they lie hidden, added unto the Bread. Such Bread is strong nourishment and far better for filling and nourishing than the common; and in eating thereof one may easily commit excess, because it notably pleaseth the Pallat by the sweetness of its taste. For so it happened on a time to me, for eating such Bread as was made up with fat Sheeps Milk I exceeded a mediocrity in my eating. But verily this is wont to happen to such as in their eating and drinking, abuse the delicate Meats and Drinks, and so fall into the hands of the Physicians, because indeed the most Diseases do arise from too much fulness and surfeiting. And therefore good reason is it that a most special regard be had to a mediocrity and temperance in all things.

XCVIII. How the Water attracting Magnet may be serviceable to Physicians.

 In my opinion a Physician may very well examine

the Urine of the Sick, and Anotamize the same, and that
more easily than by an extended bare aspect or looking
on only, if viz, he separate and take away therefrom a
part of the water by the Magnet. For by this means he
will discern a great difference betwixt the one and the
other part, and know the causes and properties of
Diseases; and that far better and more certainly than by
a bare outside view, after the Gallenical fashion, or by
the weight, and by Distillation according to the custom
of PARACELSUS and TURNHEISER; But I leave it to every
ones pleasure to enquire which of these three ways is to
be preferred.

And now follows by way of Corrolary or Surplusage,
a description of certain most excellent Medicines to be
administered, for the curing of most grievous Diseases
both in Men and Beasts.

XCIX. The Cure of the Stone in the Reins and Bladder,
and likewise of the Gout.

The Stone of the Reins and Bladder, and the Gout,
are judged to be the most grievous Diseases, and in very
deed they are most grievous Sicknesses, but especially
the Gout, which being various and manifold afflicts the
Body of Man with most grievous Pains. Now for the cure
and removal of these Diseases, I will prescribe a certain and safe Medicament, easily preparable, and of
small charge.

Take one or two pounds of white Tartar, and pour
upon every pound beaten into Powder about some eight or
ten pound of common water, which set over the Coals in a

well glazed earthen Pot, and boil it so long, till all the Tartar shall be dissolved by the water, which you may try if done or no with a wooden Spoon, putting it to the bottom and seeing if there be any left undissolved. In the boiling you must very diligently take off the Scum with a wooden Scuinmer, that so there may remain no impurity. After that all the Tartar is dissolved and that there appears no more Scum, evaporate the water so long till a thin skin appears at the top. Then take off the Pot from the Coals, and set it in some cold place, and leave it there unstirred for a day, and there will stick on to the sides of the Pot, delicate Crystals like a Dye, having a Cubical form. PARACELSUS calls this mundified Tartar LUDUS, and that very properly, and without doubt he did so, because it gets (after its purification the shape of the square Dice. Out of this pure and Cube-like Tartar is prepared an universal Medicine against all tartarous Diseases, as follows.

If you have one pound of this pure and Cubical Tartar, reduce one pound of Crude Tartar into a white Salt by Calcination; the which you are to dissolve with so much common water is as necessary to its dissolution; filter the dissolved Salt through Cap Paper, that you may have your sharp LIXIVIUM freed of all its Faeces. Pour this LIXIVIUM into the glazed Pot wherein your pound of the said purified Tartar is, and boil it accurately therewithall; in which boiling the Tartar will be easily dissolved by the LIXIVIUM, and be turned with the same into a ruddish coloured juice; though that your LIXIVIUM and Tartar had each of them a white and clear colour. The reason is this, because the Tartar is as yet defiled with many hidden and black Faeces, and doth at length after its solution with the LIXIVIUM render them visible and manifest. Pass this muddish solution through a filter, and it will be a yellow liquor, and leave many Faeces in the Philter, good for nothing but to be thrown away, for they are of no virtue more. Verily 'tis a thing worth the admiring, that there should yet be so many Faeces left in so well purified a Tartar. This liquor being thus prepared is very profitable for the taking away and curing of all kinds of Tartarous Diseases, by being daily used, or however, it doth at least strongly tame their violence, but you must first purge the Body by Antimonial Medicaments, one of which we will presently shew you.

N. B. This Medicainent will be yet far more noble, if all the humidity be vapoured away and the reddish Salt that is left be dissolved in good spirit of Wine and filtered, and the said spirit of Wine be again separated therefrom by a gentle Distillation. For so by this second solution, there will be severed yet more Faeces and the salt it self will get a yet greater purity.

This Salt may be safely used as a most precious Treasure against all the abovesaid tartareous Diseases; For it expells Urine, and drives out all the impurities out of the Reins and Bladder, and hinders the gathering together, and generation of Sand or Stones in those Members.

But if there be already Stones generated, and that

they be not too hard, it consumes them by little and little, and carrys them off; provided that Antimonial purges be (as we said but not) afore used to purge the Body with.

I have in these few words taught thee how the LUDUS, that is, the Dye like figured Tartar is changed by its own proper liquor Alkahest, or its own Alkalizate Salt into a Medicine resisting all Tartareous Diseases. The Dose thereof is a Scruple in Wine, Ale, or other Vehicles, oftentimes every day, or twice at the least, viz. Morning and Night, for such as are fifteen or twenty years old and upwards, and they must fast after the taking of the medicament, for some due time.

Such as are younger, from three, four, to ten, or twelve may take at one time, three, four, six, eight, or ten grains, according as they are older or younger. This so excellent a medicament have I described for the benefit of mankind, nor is there as far as I know, a better, though it seems to arise of so vile a Parentage, and be so mean Suffer not thy self to be affrighted by any one, but use the same boldly, whensoever necessity requires; and firmly believe me that thou wilt not find a better, I do not deceive thee; and the truth hereof will be demonstrated by its use.

This is a quick and wonderfull purification of Tartar, and a changing it into a sweet Salt, which is neither sweet nor sour, but a midling taste 'twixt both, and it gets a middle nature, from the Acid and the Alkalizate Tartar. Now follows the Antimonial Purge.

C. An universal Antimonial Purge to be used in all grievous Diseases, with a very happy success.

Take of Crude Antimony, Tartar, and Niter, of each alike, Powder them each apart, commix the Powders, being mixt, put them in a meltirig Pot or Crucible, and kindle them with a live Coal, that by this kindling they may flame up, and go into a ruddish kind of coloured mass. Your Pot being yet hot set it into your Wind Furnace, and melt it, that all your matter may flow in the Crucible like water, then pour it out into your Cone, and being cold take it out, and separate the REGULUS therefrom, and lay it by for other uses, because 'tis not serviceable for the operation here minded. Now out of one pound of Antimony, you'll have eight Lots, or four ounces of REGULUS, so that of your one pound you will get a REGULUS of four ounces or the fourth part of the pound. The SCORIA'S which will be of a reddish colour and of a fiery taste upon the Tongue must be again melted in the same Pot they were melted in, if it be whole, or in some new Pot, and when they flow, put a live Coal into the Pot, The Salt—peter will seize upon the Coal, and being occupied about corroding the same, will let fall the remainder of the REGULUS it as yet held up. Then the matters being poured out into your Cone, and cool, strike off the REGULUS at the bottom with the stroke of an hammer, and beat the SCORIA'S which will be of a red colour and fiery taste, into Powder, and being thus poudered let the Salts be extracted (or dissolved) in the heat with common fair water; the which holding in them the most pure Sulphur of Antimony do turn the water into a red LIXIVIUM, in which is hidden the Medicine that we seek after; and is to be gotten thence by the following way. For after that the Sulphur is dissolved, by dissolving all that will be separated by the Salts or Lye, the Reliques or Remainder are good for nothing.

Having so done, dissolve white and purified Tartar in fair water, in some glazed Pot, and thou shalt have an acid Solution; being thus hot as it is (for when 'tis cold the Tartar will again shoot in it) pour it by little and little into the Antimonial LIXIVIUM, and it will debilitate the same, so that the Sulphur of the Antimony will fall down to the bottom in the form of a yellowish or reddish Powder. When all the Sulphur is settled, separate the clear Water of the Salt, from the Sulphur, by canting it off; then pour on some warm Water and wash it so often till all the Salt be gone off. Then philter it, that all the Water may be separated and the Powder stay in the Philter, which you shall put upon new and dry Cap-paper thereby to remove all the wateriness, and then dry it in the heat of the Sun. This is that Universal purging Medicament which drives out all malignant humours by all the Emunctories, viz, by Vomit, Stool, Sweat, Urine, and Spittle, out of the whole Body; and that very safely if warily admiriistred, and the Dose thereof not too much encreased; in which case even the Galenical medicines themselves do hurt, if their due Dose be exceeded, Hereupon it is better that there be used at the beginning rather a lesser Dose than too much, that so no errour may be committed and a safe trial may be made, how much the strength of the Patient will bear or not bear. And albeit that the Dose of this medicament should be given in so small a Quantity as not to work, or have any visible Operation at all, yet nevertheless it well performs its Office, and drives out the Distempers, but yet more slowly than when 'tis adniinistred in a due Dose, such as may give about one, two, or three Stools, And to such as are strong and youthfull Persons, the Dose may be given in such a Quantity as to cause Vomit, in such I say as can brook vomiting. The usual Dose for those that are above fifteen Years of Age is, one, two, three, four, or five Grains, according as they are older or younger. To Infants and such as are a little elder an eighth part, a quarter, or half a Grain even to an whole Grain, may be administred, with regard had to the Age and the Disease, This medicine is of good use in all kinds of Diseases, and in all kinds of Men, (and Women) save only Women with Child; and to them you must administer either none at all, or at least wise so little as to be sure it provokes not to any inclination of vomiting. And the like is to be observed by such Persons who are so weak, as that they clearly want strength requisite for this Operation (of vomiting). But yet it may be used even to new born Infants, viz. for Convulsion, Fits with which a great many of them are wont to be snatched away without any remedy. The Dose must not exceed the bigness of a Rape-seed, and it must be administred in the Mothers milk. But if the Infants are grown already up to be somewhat stronger and are above half a Year old, the

Dose of the said medicament is to be a little encreased, that it may operate visible, and so cast forth all the malignant humours out of their Bodies; and they themselves may not be afterwards afflicted with the Small—pox, and other such like Diseases as Infants are subject unto. Verily, all the Children whom I have given this medicainent unto, have not hitherto tasted of those Diseases; the which truth the Parents of such Infants and others can confirm by their Testimonies. But especially this medicament is a most present remedy against the Epilepsie both in young and old, and a most certain Secret in the Plague and all Feavers; and a most excellent purge in the Gout, Leoprosie, French pox and other most grievous Diseases; and likewise in external new Wounds, in Fistula's and old Ulcers, what Name soever they are called by, if it be but used inwardly to purge them. Briefly, this medicament hath scarce its fellow, so that we have no reason at all to regard such men as out of meer ignorance, oppose and slander Antimony in their learned howlings, and say that it is nothing else but Poison, and therefore no ways profitable. But let no body believe their barking, but first try the same, and he will clearly find the contrary.

This indeed I readily confess that Antimony is of no use in the body of man, nor can it be, afore it is prepared; but after preparation I do boldly affirm it to be a most admirable medicine.

Concerning which, read but the chiefest of the Ancient Philosophers, especially BASIL VALENTINE, who in

the honour of Antimony wrote the Triumphant Chariot
(thereof), Do but see what PARACELSUS, ALEXANDER
SUCHTEN, and several others have recorded in writing,
concerning the incomparable Virtues and Power of
Antimony; by all whose Writings is clearly evinced, what
things lie hidden in Antimony, I forbear to speak of
other Physicians as well modern as ancient. For it
possesseth the Virtues and Powers of all Virtues and
Powers concentrated. What need many words? Let it be
brought but to the Test, and it will most apparently
testifie by its trial, what Treasures it carries in its
bosom, The things that I speak of here proceed out of
meer pity and love to my Neighbour, that I might help
his sickness and miseries. A very small Portion of the
said medicament may be able not only to preserve a whole
Family for a whole Year free from all the Diseases that
might befall them, but also to rid them of them, So
likewise it sends packing all the sicknesses of beasts
by its effectual Virtue, I was willing from a faithfull,
and good mind to bestow this comfort and help upon mankind, against all incident Diseases, whether external or
internal: In the third following Century shall be
described more very excellent medicines against all
kinds of Diseases, which may be made use of by such as
without cause are afraid of Antimony.

And as touching that Salt which is made by the
mixtion of the acid Water of the Tartar, and the
LIXIVIUM made of the Tartar caldried; it is not
inferiour as to its excellency and eminency to the but
now spoken of Sulphur of the Antimony it self, herein

only is the difference, that there is to be administred a bigger Dose thereof. And therefore it must not be thrown away, but after that the Sulphur is separated, the insipid Water is to be evaporated, and there will then remain a yellow Salt behind, which has even yet in it no small Portion of the Antimonial Sulphur; and therefore 'tis in a manner better than the Sulphur it self, because of its own peculiar Nature, which by the discharging its own office (or proper work) may be even per se reckoned up amongst those most excellent medicines which strongly resist all Tartareous diseases.

For that reason therefore do I commend this Salt most highly to all such as are burdened with grievous Diseases. The preparation thereof is altogether easie, for it is not made of any chargeable matters, but mean ones, and therefore not without cause to be highly esteemed of.

The dose of this so excellent a Salt is to be encreased or diminished according to the Quality of the Persons and the Diseases. To such as are of ripe Age, one Scruple or somewhat more is sufficient. To Infants, and to such as are a little older, from one to twelve Grains may be given, regard being had to the difference of their Years. It gently purgeth the belly, without any kind of Danger, it draws out all evil humours, and especially it helps the gouty and stony Persons with a most wisht for Easement.

The weight of those two contrary Salts, viz, of the fixt Salt of Tartar by which the Sulphur is extracted out of the Antimony, and of the common and acid Tartar dissolved in Water, and which precipitates the Sulphur of the Antimony out of the LIXIVIUM, cannot be certainly defined and limited. For according to the greater or lesser Quantity of the LIXIVIUM, is required more or less of the Tartar water to be poured upon the LIXIVIUM, that so being mortified it may let go that Sulphur of Antimony it holds up in it self. The LIXIVIUM it self will shew you if you have not poured on Water enough of the Tartar by its being not yet freed of all the Sulphur, and that there is more Water of Tartar required to allay all its Acrimony that all the whole Sulphur may be turned out. A bigger Quantity of the Water of Tartar poured on the LIXIVIUM (than just enough) doth not spoil it; it takes in as much of the dissolved Tartar as it can, and what is overplus remains an acid Tartar, and is not changed in its Nature, But whatever of it is dissolved in the LIXIVIUM is no more Tartar, because it becomes a midling Salt of the two, neither acid nor sweet, but partaking of both Natures, and dissolves in cold water, which the acid Tartar will not do.

This Salt therefore is able to perform great matters in medicine, and not only in medicine but in Alchemy too, and in other Arts can it exhibit abundance of riches; concerning which thing, more shall be spoken in another place.

Thus finish I now this my second Century, wherein I have not only abundantly supplied those things which be reason of the overmuch haste, I could not insert in the APPENDIX to the Fifth Part of the Prosperity of

GERMANY, but have withall, laid open some part of the Use of my SAL MIRABILIS, as much as the shortness of my time would give me leave to do.

If by the Grace of God I have a yet longer Life vouchsafed me, I will about half a Year hence, bring so great a benefit not only unto my own Country, but perhaps even to the whole Christian World, as ever they received from any man, in so much that the World shall seem as if 'twere new, and so for the present I rest and make an End.

THE COMPLETE WORKS
OF

RUDOLPH
GLAUBER

trans: Chris. Packe

RAMS
1983

THE CENTURYS

THIRD

THE

THE THIRD CENTURY

OF GLAUBER'S

Wealthy Store—house of Treasures.

Wherein many Profitable Chymical Secrets are discovered. Faithfully translated out of the High-Dutch of the Authour.

Courteous Reader,

Having some Years since begun to communicate to the World my manifold profitable Inventions in Centuries, but of late been hindred by sickness and other impediments from contuning the same; yet now being sollicited thereto by many Lovers of Art, I could do no less than to endeavour to give them some satisfaction by the publishing of these; and withall assuring them, that, in case God be pleased to continue my Life (notwithstanding that because of great Age and Sickness I am fain to keep my bed) I intend to compleat the rest of my promised Centuries, desiring the kind Reader in the mean time to accept of these three, and to pardon the confused manner of writing them, having for want of leisure, set them down as I found them in my Notes, being chiefly the occasional discoveries and inventions during my Chymical Labours. Neither would I have the Reader offended that in some places I break off so abruptly, especially where I am speaking of the matter which ADAM brought with him out of Paradise, for whatever may be wanting in this fifth Century shall God willing be supplied in the sixth: I also desire the Reader not to be moved by the Calumnies of any Envious ignorant Persons, to think that the things here set down (being most of them new and unheard of inventions) are mere Fables and invented matters, and no real experimented Truths, but rather remit the verification of them to time and his own Experience, which will not fail to satisfie him of the Truth of the Particulars herein contained. Farewell.

THE THIRD CENTURY.

1. To wash common Tartar Snow white in a few hours time, and reduce it to a pleasant Salt which dissolves in cold Water, and wherewith of Sugar, Honey, or any sweet Fruits at all times, yea all hours of the day, and in all places Liquors may be prepared like to Wine in Taste, smell, colour, strength and virtue, and of which afterwards good Brandy and Vinegar may be made with great profit.

2. To purifie common Salt in great quantity, in one days time, so as to become very white, pure and transparent and of a pleasant Taste, shooting into cubical Crystals fit for the Tables of great Persons, its taste being very agreeable, and the meat seasoned with it much more wholesome than that which is drest with the common Salt. SEE THE TREATISE OF THE NATURE OF SALTS.

3. A secret to preserve all sorts of Wine, and make them durable, whether of Grapes, Sugar, Honey, Apples, Pears, Quinces, Figs, Plums, Cherries, Malt, Wheat, & etc. and is of great use to a House keeper.

4. Any of the forementioned Wines may with ease be turned into very good Vinegar, not inferiour to that which is made of French or Rhenish Wine. SEE MY VEGETABLE WORK.

5. To make good SAL ARMONIACK of several contemptible matters which are trod under foot and cast out on the Dunghill very easily and in great quantities, so as one Man every day may prepare one hundred pound weight of it with ten shillings charges. SEE MY TREATISE OF THE MINERAL SQUILL IN ORDER TO LONG LIFE.

6. A secret water wherewith in an hours time the yellow colour in Diamonds may be drawn from them, which makes them ten times more worth than they were before. SEE MY TREATISE OF THE DIVINE CHARACTER.

7. In like manner may the dark red colour of Granates be extracted, leaving them only so much colour as makes them like Rubies. For Granates and Rubies resemble each other in their bodies and colour, the only difference between them being, that the Granates abound with too much colour, which makes them less valued, when therefore some part of their colour is extracted from them, they do in virtue, hardness and beauty, equal Rubies, one Karat of which is more worth than ten pound of Granates, so as this extraction must be very gainfull to him that is Master of it. SEE MY THIRD APPENDIX TO THE SEVENTH PART OF MY PHARMACOPOEA SPAGYRICA.

8. In like mamier also may be extracted the colours of blue Saphyrs, yellow Jacinths, Topaces, and Purple Amethysts, by which means they become white as Diamonds, and when brought to the same degree of hardness are every whit as valuable as they. SEE MY THIRD APPENDIX AS BEFORE.

9. In a moments time to rob SOL of its colour and make it white as Silver. SEE MY TREATISE OF THE SEAL OF GOD.

10. To separate from MARS and VENUS when dissolved in Water as well as from any other Vitriol, by means of an artificial Precipitation, their hidden spiritual SOL or Tincture, and that in a moment; a thing of great use in Physick, as well as in the transmutation of Metals. SEE THE SECOND APPENDIX.

11. To extract SOL out of Sand and Stones with great ease and little charges, which by precipitation is afterwards separated from the dissolvent, retaining its former strength, and may be made use of again for the like extraction. SEE THE PROSPERITY OF GERMANY THE SEVENTH PART, OR NOVUM LUMEN CHYMICUM.

12. To extract SOL from LUNA with a small quantity of dissolvent, which, after precipitation of the SOL, remains in its full strength, and may be used as before to the great gain of the Artist. SEE GLAUBER'S LABORATORY, AND PROSPERITY OF GERMANY, 7th. PART.

13. In one days time to prepare a particular, whereof one part will tinge three parts of VENUS into LUNE. N. B. This Tincture is a white Stone which being placed in a fit Furnace, and a due fire administred,

within few days the whiteness will be changed to a
yellow colour, and that into a fixed red, whereof one
part being cast upon four parts of LUNE in Flux, exalts
it so far that in the separation it gives a fourth part
of SOL. Which sudden fixation is performed by the proper
Agent of the matter which is white of it self, and yet
affords a red Tincture, when handled, as is here set
down, SEE MY TREATISE OF THE SECRET FIRE OF ARTEPHIUS.

14. In a short time to prepare a particular
Tincture of a red subject, which exalts Silver to that
degree, as to yield much SOL in the separation. SEE MY
EXPLICATION UPON PONTANUS HIS LETTER.

15. A good graduating water which being digested
with LUNE, makes it yield much SOL in separation, SEE
THE TREATISE CONCERNING THE MOST SECRET NATURAL SAL
ARMONIACK EVERYWHERE TO BE FOUND.

16. Another graduating water in which Mercury
being digested, becomes coagulated into SOL and LUNA.
SEE THE FIERY ALKAHEST.

17. Another fixing water, which being once or
twice abstracted from Mercury makes it lose its property
of making SOI, and VENUS white, and on the contrary
gilds LUNA when rubbed upon it. I have as yet carried
this experiment no further, but am of opinion that if
Mercury were long enough digested in the same, it would
turn the Mercury into Tincture, coagulating and fixing
each common MERCURY into SOL. SEE HASTECAL.

18. A volatilizing Water which being abstracted
from SOL highly exalts its natural colour, and carries
it over the Helm, which done it is no more common SOL,

but may in a short time be fixed into a transparent red Carbuncle. SEE MY THIRD APPENDIX, & etc. concerning the Griffins Claws, and Eagles Wings.

19. A water of like nature that volatilizeth all fixt matters, wherewith in one single Distillation, the Tincture or Soul may be extracted from MARS, VENUS, and all coloured Stones, and carried over the Helm; which Tinctures afterwards with one rectification are highly purified, and have their Medicinal and tinging virtue doubled, which exalted and multiplied Tinctures, notwithstanding their great volatility may within twenty four hours time be concentrated, by means of a secret Magnet, and fixed into a Stone, penetrating all compact Bodies, with which incredible things may be done in Alchemy and Physick. SEE MY THIRD APPENDIX, & ETC.

20. To prepare a Salt in an hours time, and without extraordinary charges, which makes all fixt matters volatile, and is of such virtue that when a little of it is joined with Spirit of Wine it makes it so strong and fiery, that it dissolves all Metals, Minerals, Animals and Vegetables, carrying their Quintessence over the Helm, and is the effecter of wonders in Physick and Alchemy; so that he who knows how to prepare and make use of this wonderfttll Salt, needs never want either bodily health, or a competent supply of maintenance. SEE MY TREATISE OF ELIAS ARTISTA IN QUARTO.

21. A wonderfull, to all Men known, but withall contemptible matter; which every where may be had for nothing, which whosoever knows, together with the use of it, needs never want, because thereby he may effect

279

whatsoever is necessary for Soul or Body. SEE MY FIRST, SECOND AND THIRD APPENDIX OF PHARMACOPOEA SPAGYRICA.

22. The manners of preparing a running Mercury out of all Minerals and Metals, and that in one days time, which joined with SOL becomes fixed into SOL. SEE MY THIRD APPENDIX.

23. How such a Mercury may be prepared in an hours time of the martial REGULUS of Antimony, without any charges to speak of, which is a true Tincture, fixing the imperfect Metals into SOL. SEE MY TREATISE CONCERNING THE SECRET FIRE OF THE MAGI.

24. A water made of a particular sort of Chalk which changeth a yellow or brown skin, into white, and which cannot be washed off with water, of valuable use for Ladies and Gentlewomen. SEE MY FOURTH CENTURY IN FOLIO.

25. A water prepared of SOL, which turns white hairs into a yellow gold colour. SEE MY LABORATORY.

26. Another water made of Silver, which tinges hair cole black, good for such as are grey haired, and endeavour to conceal their Age. SEE GLAUBER CONCENTRATE.

27. A water made of SOL, which colours the hair and skin of Nan, as also the bones and horns of Beast, and feathers of Birds, of a fair lasting Purple. SEE GLAUBER'S LABORATORY IN QUARTO.

28. A water into which when any Metal is put, it begins to grow within twenty four hours time in the form of Plants and Trees, each Metal according to its inmost colour and property, which Metalline Vegetations are called Philosophical Trees, both pleasant to the Eye and

of good use. VIDE MY FOURTH CENTURY.

29. A water made of Sand and Flints, having the property of changing Wood that is laid in it, in a short time into hard Stone of several colours according to the pleasure of the Artist.

30. A dry water, or rather Stone, upon which if any volatile saline Spirit be poured and set in the Sun, it presently sucks in the volatile mineral Spirit and in one days time makes it so fixt that it may be made red hot in the Fire, without any evaporation. SEE MY UNIVERSAL COAGULATOR.

31. By this means also may the combustible stinking Sulphur, the greatest enemy of Metals be fixed, which afterwards being cast upon the imperfect Metals in Flux, doth meliorate them, and make them afford SOL and LUNA on the Cupel with profit. SEE MY SECOND APPENDIX.

32. In like manner may Antimony without any loss of weight be fixed, so as no more to cause vomiting, but casts all evil out of the Body insensibly by sweat, restoring health, and renewing youth. SEE MY PROPER AGENT.

33. In the same manner may Orpiment be fixed, so as no longer to be a Poison, but a Meliorator of imperfect Metals.

34. Likewise also may Arsenjckbe fixed within two or three days time, so as it may safely be taken inwardly, being an excellent Diaphoretick for the cure of Diseases, and good to exalt Metals, so that in separation they afford Gold and LUNE. SEE MY PROPER AGENT.

35. Much after the same manner may Mercury, without any considerable loss of weight (though with longer time, and more patience) be fixed, so as to suffer himself to be melted and hammered like any other Metal, and on the Cupel leaves SOL and LUNA. SEE MY TREATISE OF THE UNIVERSAL COAGULATOR.

36. In like manner may the martial REGULUS of Antimony be fixed into a tinging Stone, that meliorates all imperfect metals.

37. A wonderful Magnet which being put into any watry Liquors or Oils, draws the water to it self leaving the Oils more pure, subtle and penetrating. SEE ELIAS ARTISTA.

38. By means of this Magnet, we can separate from the highest rectified Spirit of Wine, one half of insipid water, which Spirit of Wine after this separation is an effecter of wonders.

39. This Spirit of Wine wnen poured on pulverised Coral and thence abstracted, brings their red Tincture over the Helm, being a wonderfull Cordial and purifier of the Blood.

40. By means of this Spirit of Wine, may the Cordial Virtue of Pearls be brought over the Helm, being of great efficacy for the recovery of sick and weak Persons.

41. This Quintessence of Wine being poured upon clean washed Egg shells, dissolves them, and distilled from them, brings over with it their Stone-breaking and dissolving virtue, and is a singular remedy in the Gout, and the Stone of Kidney or Bladder.

42. The same also dissolves the LAPIS LYNCIS and JUDAICUS, as also Crabs Eyes so called, and other Stones found in Fishes, carrying their virtues with it over the Helm. SEE MY FIFTH CENTURY.

43. The same Spirit of Wine dissolves, extracts and brings over the Helm, the inmost virtues of all Animals and Vegetables carrying them over the Helm, whence incomparable medicaments may be prepared.

44. Black Snails such as are found in MAY on the Grass dissolved in the same, and brought over the Helm, and duely exhibited to those that have the Gout or Stone, carries off all tartareous slimy matter from the Kidneys, Bladder, and other parts of the Body by Seige and Urin.

45. Aloes, Saffron and Myrrh dissolved in the same, and their Tinctures carried over the Helm, affords an excellent ELIXIR PROPRIETATIS, very conducive to long life. SEE PARACELSUS CONCERNING ELIXIRS.

46. Canthandes dissolved therewith, and brought over, are a powerfull Diaphoretick, above all others, cleansing the Kidneys and Bladder, but ought to be heedfully used, because it is a vehement Medicine, which being overdosed will hurt the Kidneys, Bladder and Ureters.

47. The Leaves of Helleboraster, extracted and brought over with the same, affords an Excellent AQUA VITAE, conducing to long life. SEE PARACELSUS.

48. Sea Squills being dissolved in the Spirit of Wine, and spiritualized by being brought over the helm, is of great use for removing of Diseases, and

maintaining of health.

49. NUX VOMICA, being first grated and then dissolved therein, and their restorating Virtue being brought over the Helm, doth wonders in restoring the decayed strength and health of man, but must be used with understanding.

50. Common Mercury dissolved and brought over with the same, is the highest Medicine against the French Pox and all venereal Diseases.

51. Mercury of Antimony prepared after the same manner, affords a Medicine against all Diseases of mankind.

52. Fixed Antimony thus extracted & brought over is a Diaphoretick curing all Diseases, and restoring the highest degree of health.

53. In like manner may many excellent Remedies be prepared, out of all Vegetables and Minerals, for restoring and preserving of health.

54. A further use of our Water attracting Magnet is this. Abstract the Oilof Tartar, Hartshorn, Amber, Soot, or that which is distilled from Smiths Coals, and the Magnet will attract all the Water and bad Smell to it self, which remains with the Magnet, and the pure clear and subtil Oil, only comes over, which Magnet being made red hot, loseth its water and stink and may be made use of as before.

55. In like manner may the Oil of Wax and Bricks, commonly called the Oil of Philosophers, be deprived of their bad Scents, and made exceedingly penetrative and pleasant.

56. So also may all Vegetable and Animal Oils distilled by retort, be purified and made pleasant.

57. Likewise all the Oils of Herbs, Seeds, Woods and Spices, which with the addition of Water are distilled by a common Still, may by distilling them from our Magnet be made much more subtil, and their sweet Smells much more strong and piercing; so that a little of these Oils set in an open Vessel, perfumes not only the Room in which it is but also the whole house, they being so volatile that without any Fire they vanish in the Air.

58. And as by means of this volatalizing fiery Spirit, the pleasant and well scented Oils of Spices may be greatly meliorated and exalted; so likewise may all stinking and poisonous Vegetables, Animals or Minerals, thereby be made much more stinking and venemous, so as their Smell alone will be sufficient to kill men, doing it with far greater expedition than any Corporal poisons whatsoever. SEE ELIAS ARTISTA.

59. All well scented Oils, may by means of our volatile saline Spirit, be purified to the highest Degree, and afterwards be reduced to a hard Body; which Body then is no common gross Body (as being a coagulated Spirit) but a clear, transparent spiritual pure Body.

60. This Labour may with profit be practised on Amber, whose Oil being by rectification made clear and transparent, and then digested with our fiery salt Spirit, becomes hard again as it was before distillation; by which means we may make pieces of Amber as big as we please, and may mix with it some small

Threads of SOL, and so shall have the old highly esteemed Stone called Chrysophoros; or else we may put into it, whilst yet it is soft, little Worms, Flies, Spiders, Pismites, or whatever else we please, which is a notable Cusiosity and shews as if they were grown there, to those that are ignorant of this Art.

61. In the same manner may the Oil of Terpentine be reduced to a hard Gum, to very good use and purpose.

62. All distilled Oils or Seeds, Woods and Spices, when by long standing, they turn yellow, red or thick, may by means of this fiery salt Spirit be again made clear, thin and transparent, when some of the said Spirit is poured on the said Oils and so distilled, some part of the Oil comes over clear and transparent, the other part remaining in the Glass, in the form of an hard Gum, in which small Insects may be inclosed as before said of the Amber.

63. Amongst all Oils these following are apt to grow thick and ropy, viz, the Oils of LIGNUM RHEDU, Oranges, Limmons, Juniper berries; those of Cloves and Cinainmon are apt to grow red. The Oils of Fennel seed, Annis, Coriander, Caroway and Cumin seed, and all other Oils distilled from Herbs and Seeds, that have hollow Stalks, and are umbelliferous, forasmuch as they abound with much volatile Salt, are apt to turn thick: If any of these be rectified with an acid saline Spirit, it immediately destroys the volatile Salt, and the Oil becomes clear and transparent, and the remaining part of the Oil becomes hard as a Gum, and is a special inward and outward Medicine.

64. And forasmuch as a fiery saline Spirit can make old and red Oils clear, thin and transparent, we may conclude, that such a Spirit is able also to volatilize and bring over by distillation those Oils which by length of time are become hard and dry in Seeds, Herbs and Woods, and cannot by maceration in water be brought over, but must by this more powerfull means be made thin and volatile, that they may afford their Oils as easily in distillation, as green Seeds and Woods are used to do.

65. Now as thick and ropy distilled Oils may be made thin, by means of saline Spirits, so there are some salt Spirits wherewith we can coagulate all thin and subtil Oils, in the form of a volatile pleasant strong scented Salt, of great use in Physick.

66. In this manner, viz, by pouring a strong saline Spirit upon them, we can distill subtil and powerfull Oils from all rosins, gums and thick juices, and afterwards reduce them again to the hardness of Amber.

Thus Mastick, Frankincense, Benzoin, Storax, Camphir, & etc. afford very pure clear and transparent Oils, which when hardened to the consistence of Amber, draw straws and other light matters to them like it.

67. In like manner also can all sulphureous Minerals be purified to the highest degree, when distilled with such a Spirit, and then may be reduced again to hard transparent clarified Bodies; and amongst the rest Antimony and Orpiment do afford in this way most powerfull and superlatively penetrating medicinal

Stones.

68. And as these fiery saline Spirits do bring over by distillation all unfixt sulphureous Subjects, and purifie them, so they do the same in fixt Sulphureous Metals, Eg, MARS and VENUS, which being thus purified may be fixed into tinging medicinal Stones.

69. They who know the art of the metallick purification and fixation, are possessours of an incomparable Treasure, forasmuch as by this means in three days time with the charge of one Crown, a true universal Medicine may be prepared, for the Bodies of men and metals, not in great quantity indeed but sufficient to assure the Possibility of it, and may afterwards be tryed in greater quantity.

70. By means of such a fiery salt Spirit fixt Chrystals, Flints, and other hard Stones may be made volatile and spiritual, and then may tinge them with what colours we please, and coagulate them again into hard transparent coloured Stones, and that of what form and fashion we please. This is a very gainfull Art, because fair transparent coloured Stones are always preferrable to SOL.

71. And as we have understood that by means of volatilizing waters, not only Vegetables and Animals, but also minerals and fixt metals may be made volatile, and their purest parts brought over the Helm, and by this means do wonders as well in Physick as Alchemy. Now through this separation of the pure part from the impure, by means of Distillation be highly to be valued, yet there is a better, easier and less chargeable way to

separate the purer parts from all Metals, Minerals, Stones, Sand, and coloured Earths containing SOL and Tincture, by means of a Magnet, which being laid in the Solutions of Metals, and extractions of Stones, within a few hours time draws to it all the spiritual as well as fixt SOL and Tincture contained in the said Solutions; so that after abstraction of the dissolvent by Distillation, we find the dis-animated dead Body, which being put aside, we separate the attracted SOL or Tincture from the Magnet, and thus obtain whatsoever good was hid in the aforesaid gross Bodies.

N. B. Though indeed this extraction of SOL and Tinctures be very easie, as hath been said, yet I shall here, for further information of the Reader, set down what ought to be observed in the extracting of SOL from each Metal, Mineral, Stone or Earth, And first fixt SOL.

72. When there is fixt SOL in Sand or Stones, we need only to pulverize them, and pour upon them AQUA REGIS wherein common Salt hath been dissolved, and let them boil together a quarter of an hour, by which means the AQUA REGIS draws the SOL out of the Sand and Stones, in which extraction if we then put the SOL attracting Magnet, it will draw the SOL to itself, which being separated from the Magnet, is melted down with fluxing Powder, the AQUA REGIS continuing good to be employed on the like occasion.

73. But when in the Sand or Stones there is no fixt but only a volatile unripe SOL, then we must put the Sand or pulverized Stones into a Glass Retort, and pour upon them of our volatilizing fiery Alkahest, and

abstract the same from the Sand or Stones, by which means it carries the SOL over with it, which hath been attracted by the golden Magnet, which being reduced will be found good and fixed SOL. The dissolvent may again be used to the same purpose.

74. But if the Stones besides the SOL, do also contain LUNA, then an AQUA FORTIS must first be poured upon them to extract the LUNE, and afterwards precipitate it by casting some common Salt into the Solution, by which addition of Salt the AQUA FORTIS is turned into an AQUA REGIS, and being poured upon the Stones it extracts the SOL also.

75. For if at first we should pour an AQUA REGIS upon these Stones, it would indeed extract the SOL, but withall so alter the LUNE, that it would be impossible afterwards to extract it with AQUA FORTIS and therefore we are to proceed in the manner above-said.

76. In like manner also we are to proceed with white, yellow and red Earths, for to extract the LUNE and SOL that is in them. And if it be a fat Earth and contains fixed SOL, it must first be made red hot to rob it of its fatness because else it would devour too much AQUA FORTIS.

77. But when the Earth contains only volatile SOL as the yellow Earth of SILESIA and the red TERRA LEMNIA SIGILLATA, it must not be made red hot, but pour some Alkahest upon it, and so bring the SOL over the Helm.

78. Yellow and red Earths do commonly contain SOL or LUNE or both, and therefore we may boldly make trial of them; for oftentimes a great treasure is shut up in

very contemptible Earth such as might serve to maintain many thousands.

79. In the same manner we may extract the fixt as well as immature SOL out of Ruddle, red Jasper and red Blood-stones, which in some parts are found in great quantities.

80. With our SOL attracting Magnet we can extract much SOL out of any common Vitriol, and after the SOL is extracted, reduce the Solution to Vitriol again, which is as good for the Dyers use as it was before.

81. With the same Magnet good SOL may be extracted out of those yellowish reddish and greenish waters which flow from some Nountains, which waters being to be had for nothing, must make this work very gainfull.

82. And in case we should not be able to meet with this sort of running waters, then we may take the Copperas Stones which are often found in Sand, but for the most part grow in fat Earth, which when they are exposed to the Air, fall in pieces, and having water poured upon them afford a good Vitriol, which easily yields the spiritual SOL it contains to our Magnet, So that an Artist can scarcely be to seek for subjects; from whence unripe volatile or fixt SOL may be had with profit.

83. N.B. But when we have a mind to bring the volatile SOL which is in red Sand, Stones and Clay, with volatilizing waters by Distillation over the Helm, it is good to add to our Eagles Wings, or volatilizing water, some of our most secret SAL ARMONIACK, by which the same is extremely strengthened and animated, so that like a

GRIFFON for strength, it carries the Man on Horseback away in the Air to his Nest.

84. N. B. This GRIFFON is the Artist, that prepares this fiery water, wherewith he seizeth as with his Talons the Man armed Cap a Pe; that is, red Stones, Sand and Earth, abounding with a martial Tincture, extracts and carries them to the young ones in his Nest, that is, provides a good maintenance for his Family thereby.

85. But because to these operations of extracting SOL volatile and fixt, and Tinctures from the subjects just now mentioned, when we work them in quantities, much AQUA REGIS, or other like waters are required, which every one hath not an opportunity to prepare for himself, and therefore must buy them, which encreaseth the charges of the operation; wherefore I am willing to teach a near way to prepare these corrosive waters.

86. Forasmuch then as we know that Vitriol is an universal acid, and the chief of all Salts, and the Spirit it yields by Distillation, much more fiery than that which is forced from other Salts, therefore we may make use of the Oil of Vitriol, for a BASIS with the help of other Salts, to prepare several sorts of saline Spirits, with small labour and charges, in manner as follows.

87. RECIPE two parts of Niter dissolved in water and one part of Oil of Vitriol, distill them in an Alembick, and you will obtain a good AQUA FORTIS to dissolve LUNE, SATURN, and MERCURY. This operation spends little Fire and comes over easily.

88. And if we dissolve one pound of Salt, and as much Niter in three pound of Water, adding thereto one pound of Oil of Vitriol, and distill it in Sand, by Alembick or Retort we shall get six pound of good and strong AQUA REGIS to dissolve SOL, VENUS, MARS and JUPITER.

89. But when we take two pound of Salt, and dissolve it in three pound of water adding one pound of Oil of Vitriol, we get five pound of good Spirit of Salt. The Salt that stays behind is called SAL MIRABILE or wonderfull Salt, because wonders may be done with it, as appears from several parts of my Writings, these Salts being of divers virtues according to the nature of those Salts that are added to the Oil of Vitriol in Distillation.

90. These Salts are commonly added to metals, and melted down with them in Crucibles, by which means they become dissolved in the dry way, which is much easier and readier than the wet way of dissolving.

91. In particular by this way we can dissolve Sulphur, which resists all corrosive saline Spirits, and remains undissolved by them.

92. Now to obtain Oil of Vitriol with ease and in great quantity, we may proceed several ways, and especially thus, by dissolving Vitriol in water, adding a contrary to it, which separates all its impurities, by which means the purified Vitriol may with a small Fire be reduced to Oil, so as one pound of Oil of Vitriol will not require above ten pound of Coals.

93. And because Oil of Sulphur is of the same

nature with Oil of Vitriol, yea is more proper for some operations than it, therefore we may make use of the same Oil of Sulphur, to prepare strong saline Spirits, especially because the same may be prepared in quantity and very compendiously, according to a particular way described by the Ancient Philosophers.

94. They have taught us the preparation of Oil of Sulphur in their wittily devised Fables, giving to this Oil the name of VENUS, whom VULCAN when come to Man's estate took to Wife; by the word VULCAN, we are to understand every combustible Sulphur, and by the word VENUS, its incombustible corrosive Oil, which for this reason probably they called VENUS, because when a drop of it falls on burning Coals, it gives forth a red smoak like to VENUS, or because this acid Oil like a wicked Woman, has sharp Teeth, and a keen deceit-full Tongue, wherewith they lay hold of Men, and lead them astray in the same manner as this Oil cleaves to, and enters a League with every metal to which it is joined, forasmuch as all metals proceed from Sulphur, and have great affinity with it, as the Woman hath with Man.

Here follows an Explication of the Poetical Fable, teaching us to make the Oil of Sulphur in quantity.

95. We read that VULCAN, that is, a combustible Sulphur, took VENUS to Wife, by which is meant the incombustible Oil of Sulphur; now whilst VULCAN was busie at his work in the Caves of the Earth, for he was a Miner and a Black-smith, VENUS betakes her self to MARS, who lies with her; now when VULCAN found his Wife in the embraces of another Man, he calls all the Gods to

his help, who appearing, casts an Iron Net over MARS and VENUS that they might not escape, but with great shame be exposed to the derision of the Gods.

96. When VULCAN is kindled in a firnace made for that purpose, that is, labours in metals, the flame carries VENUS, i. e, the Oil of Sulphur into MARS his lodging, that is, into the Recipient, which must be filled with Steel Wire, where she commits Adultery with MARS, that is, begins to dissolve the same, and produceth a Vitriol, which when exposed to the moist Air, becomes resolved into a clear sweet water, which from the Steel Wire runs into the Vessel appointed to receive it, which sweet water is nothing else but AROSTRUS the Son of MARS and VENUS, or the martial Oil of Sulphur, or to speak more plainly, the Vitriol of MARS.

97. This Vitriol of MARS is called by the Philosophers the juice or water of the Birch Tree, and of which they say the Philosophers Stone may be prepared, for many of them have writ concerning it, and pointed to this sweet Iron Juice, which they have termed Birch Tree Waters, because of the likeness it has therewith, for when in the Spring time we make a hole with a Piercer, reaching to the Pith of the Tree, and thrust into it a Quill or Faucet, much sweet water proceeds from it, which some counted very good against the Stone in the Kidneys and Bladder.

98. This Vitriol also is VIRGIL'S ARBOR OPACA, or dark Tree, whose Branches he declares to be easily flexible; now what is more flexible than Iron, or Copper

Wire, which are therefore compared to the Birch, which is a very juicy Tree, and hath very flexible branches.

99. That this shady Tree is the fittest, according to VIRGIL'S Doctrine, to break off one golden Twig after another is also certain, for I have tried it divers ways and found it to be true, that it affords its golden Twigs very freely to him that knows how to handle PROSPERPINA.

100. But if the Artist be acquainted with the use of our SOL attracting Magnet, he may then with ease and more profit, without Distillation or Fire, break off the said Twigs, than he could do with the help of PROSERPINA.

And thus I conclude this third Century, the fourth and fifth follows, which will be found of better use than this.

THE COMPLETE WORKS

OF

RUDOLPH
GLAUBER

trans; Chris. Packe

RAMS
1983

THE CENTURYS

FOURTH

THE FOURTH CENTURY

To extract the SOL that is in Granates.

1. If Granates be melted with Glass, that contains LUNE containing Glass will unite it self with the SOL that is in the Granates, which with an AQUA REGIA may thus be afterwards separated.

To make a good MERCURY of Wine.

2. RECIPE Crude Tartar and pour upon it a Lye of Salt of Tartar, distill in an Alembick, and you'll have a strong Mercury of Wine, which is a much stronger and more fiery Spirit, than Spirit of Urine, especially if some SAL ARMONIACK be added to it.

To make a Mercury of Metals.

3. And if before we dissolve a Metal or Mineral in the aforesaid lye of Salt of Tartar, or in the Crude Tartar, before that both these contraries be put together, then the volatile Spirit of Tartar will bring over the Tincture or Mercury of the said metal or mineral in the form of a subtile spirit. In this manner we may drive the Mercury of all minerals and metals over the Helm.

What the Soul of the greater and lesser World is.

4. PLATO called common Salt the Soul of the great World, and if so, then Salt of Tartar may very well be accounted the Soul of the lesser World; for whatsoever is in the Macrocosm, the same is also in the Microcosm. All superfluities of Nature afford a volatile Salt.

5. For the Salt of Hartshorn of Hair, Soot, Blood, and of the Seeds of Mustard, Cresses and Scurvygrass, & etc. are much of the same Nature as SALT of Tartar.

A Good Bath.

6. Common water sprinkled upon red hot Flints or Pebbles that are found in running Streams, affords an hot Vapour, which by reason of the subtil Sulphur of the stones it carries up with it, is very penetrating, so that in this manner without any other Fire we may prepare an hot dry Bath, very available to cure many Diseases by sweating, the great virtue of it chiefly proceeding from the sulphureous Spirit of the Stones.

To separate SOL from LUNA by fluxing in a Crucible.

7. When we have a mind to separate SOL from LUNA by means of Sulphur we need not make use of granulated or filed SATURN for precipitation, but instead thereof we may make use of Antimony, because the granulated or filed SATURN is made hard and influxi'ble by the

Sulphur; neither shall we make use of common VENUS for precipitation, as ERKER teaches, but such a VENUS as hath been made friable with Arsenick or Orpiment, by which means we shall get more SOL than without Arsenick, because Arsenick and Orpiment contain much volatile SOL, which in this Operation becomes fixed with the LUNA.

To recover the SOL and LUNA which is got into the Pores of the Crucible.

8. The Crucible must be beaten into fine Powder, and put into a reverberatory fire, stirring it continually by which means the Antimony and Sulphur vapour away, and the SOL and LUNA remains with the Earth, which, with strong Waters may be separated.

Another way to perform the same.

9. Or we may add filings of SATURN to the powder of the Crucibles, and give them a strong heat, by which means the SATURN will take in the SOL and LUNA. N. B. But the Separation with strong Waters is the more easie way of the two.

To extract the Colour from SOL.

10. VENUS, JUPITER and REGULUS MARTIS melted into a mass with SOL, and VENUS, the JUPITER and REGULUS MARTIS afterwards separated from the SOL by NITER, then melt other metals as before with the SOL, and separate

them as before with Niter, which must be continued till the SOL have lost his Colour.

11. The dross being afterwards melted in a Crucible, and a small quantity of Coles made of Blood cast upon it, the extracted Tincture of SOL will separate it self from the dross, falling to the bottom like a REGULUS. N. B. The metals VENUS, JUPITER and the martial REGULUS may be separated from the dross only with common wood Coal.

12. Niter fixed by the REGULUS of Antimony, and distilled with SAL ARMONIACK, gives an excellent MERCURY of metals, which hath a scent like musk.

To extract SOL out of Stones.

13. If we take Sand or Stones that contain SOL, and melt them with lead ashes into Glass, and reduce them again with fixed Salt, then by cupelling this Lead ashes, and the reduced Sand or Stones several times, we shall have the SOL that was in the Stones.

To extract SOL from Stones.

14. The black or brown Pebbles found in Brooks, and which break smooth like Glass, being mixed with the best Eagles wings and distilled by retort, yield much SOL.

A Tincture from Metals.

15. JUPITER is the highest Regent over the upper Constellations, SOL gives to all Stars their Light, MARS rules upon Earth, and SATURN in the Earth, and of these four an universal Tincture may be prepared.

16. MARS and SATURN in particular yields great riches, when being reduced to Glass they are several times driven through a Reverberatory, according to that of the Ancient Philosophers; by SATURN and MARS, fire and art, great wealth may be found.

That there is a renovating virtue hid in Spiders.

17. Spiders renew themselves every month by casting their skins, wherefore a medicine prepared of them by the Flame of Spirit of Wine renews man.

18. It is also to be observed that all Birds, especially those that feed upon Flies and Worms, when they are sick, cure themselves by eating Spiders.

Secrets of Serpents.

19. All sorts of Serpents renovate themselves once a year by casting their skins, wherefore if we extract them with Spirit of Wine, and correct theni by burning away the Spirit of Wine, they yield an assured medicine against all Poison, and renews man.

20. REGULUS of Antimony being duely fixed with SOL, tingeth as well in the wet as dry way.

SOL and Sulphur yield a Tincture.

21. Common combustible Sulphur cannot join with the incombustible Sulphur of SOL, without such a medium as partakes of both their Natures, viz. Antimony; when by this means the combustible Sulphur is fixed by the incombustible Sulphur of SOL, the SOL, gives it ingress into imperfect metals to tinge them.

To make SOL red.

22. If the Blood of the Lion be digested with Tartar and AQUA FORTIS, this purple Colour will be changed into a red, and separate it self from the Salt, falling in form of red powder to the bottom, and is a most excellent Colour for Painters.

To make PURPURISSA, or a Paint to make the Face look ruddy.

23. Dissolve SOL and JUPITER in AQUA REGIS, digest and edulcorate with Water, yields an excellent paint for Women. N. B. But a little Oil of Talk ought to be added to it.

An Experiment upon PURPURISSA or the Blood of the Lyon.

24. When we digest or boil the Blood of the Lion so long till the red Colour becomes as white as milk, and then pour upon it as much Water, as has been evaporated during the boiling, this milk will be changed

again to Blood.

25.   Of the blue Paint called Smalt, by means of
Salt of Tartar may a most excellent Paint be made for
Limners, not inferiour to Ultermarine.

26.   Of MERCURY, JUPITER, Sulphur and SAL
ARMONIACK is made the Paint called AURUM MUSICUM.

A Cementation that graduates VENUS into SOL.

27.   RECIPE Vitriol calcined to Redness, mix it
with Salt and Coal dust, lay this with thin Copper
Plates STRATUM SUPER STRATUM, put them into a Fire that
may keep the Plates of VENUS red hot for six hours
without melting them, by which means the SOL in the
VENUS, will be encreased; if we repeat this Cementation
several times till the VENUS be of a golden colour, the
gain will be much greater.

N.B. The cause of this melioration is, because the
Coal Dust hinders the corrosive Spirits of the Vitriol
and Salt from corroding the VENUS, and therefore only
penetrates and graduates the same.

To make all Corrosives sweet.

28.   Vitriol distilled with Salt yields a
Corrosive Spirit, but if Coal Dust be mixed with them,
they give a sweet Spirit, which graduates LUNE into SOL
when digested therein.

29.   RECIPE, CALX of JUPITER mixed with MERCURY of
LUNE, and therewith Cement plates of VENUS, by which

means the saline Spirits introduce the white Sulphur of JUPITER into the VENUS, and change it into LUNE containing much SOL.

A sweet graduating Spirit, usefull to the Melioration of Metals.

30.   RECIPE, Coal Dust, mix them with SAL MIRABILIS, and. distill by retort, and you will get a sweet graduating Spirit, exalting some Metals to SOL.

The Philosophical Work.

31.   The Father of all things is the warm Son, their Mother is the moist Moon, the Earth is the Womb, the Wind carries the Seed through the Suns drMng into the Womb the Earth, which foments, and at last brings forth the Child.

Sulphur is the Father of all Metals.

32.   The Central Fire in the Earth mounts upward continually into the hollow places of the Earth, and meeting with water or moisture, cleaves to it and makes Stones, as also all Metals and Minerals of different natures and properties, according as the water is pure or impure.

Sulphur is the universal Coagulator.

33.   A sulphureous vapour is that which coagulates Mercury, as well in Vegetables and Animals as Minerals.

Demogorgon the Grand-father of all things.

34.   The Central Fire in the great World keeps it in continual motion, and causeth the growth of all things as well upon, as under the Earth, being the Governour of the great World.

The Vital Spirit, or radical moisture, is the Life and Growth of all Men.

35.   As the great World is governed and maintained by the Demo-gorgon or Central Fire, so Man the little World is governed, and maintained in continual motion and growth, by the vital Spirit seated in his heart.

36.   Fire is the Father of all things, Water the Mother, the Earth is the Womb, the Wind or Air drives the Fire, being the universal Agent, into the Water, being the universal Patient, in order to bring forth Fruit. SEE MY TREATISE OF THE DIVINE CHARACTER.

37.   Man, Beasts, Fish and Fowl, and all that lives and grows, draw their life from the Air, only the SALAMANDER draws his life, and his Body from the Fire, wherefore also in power and strength he excells all living Creatures.

38.   The secret Fire of the CHALDEANS, which at all times draws Fire out of the Air, wherewith the Jewish Priests kindled their Sacrifice, as may be seen

in the MACCABEES, is made of Steel, Niter and Sulphur.

39.   When we abstract an AQUA REGIA wherein SOL is dissolved from the Butter of Antimony, the Soul of SOL and Antimony comes Blood-red over the helm, which poured upon a Solution of LUNE, the LUNE falls to the bottom, and draws the Tincture of SOL and Antimony to its self out of the Water, and the LUNE by this means becomes red, and is a Tincture and Universal medicine for humane and metallick Bodies.

N. B. The remainder of the SOL and Antimony that did not come over is wholly fixt and a good Diaphoretick, thus the Souls of the dead, i.e. of SOL and LUNE are brought up from Hell. SEE NUISEMENT DE SPIRITUS & SALE MUNDI.

40.   When in the manner now said with the Butter of Antimony, we bring over, the Soul of MARS we get a much higher Tincture than from that of SOL, and in coming over becomes wholly fixt, SEE MY TREATISE DE 3 PRINCIPITS METALLORUM.

41.   In like manner may from the Butter of Arsenick and LUNE a white Tincture be brought over the helm, tinging VENUS, MARS, JUPITER and SATURN into LUNE. N. B. These Tinctures in coming over are fixed by PLATO'S Stygian Water, so as to need no further fixation.

42.   But if we precipitate these Tinctures of MARS and Antimony with the Solution of SOL, and then edulcorate and dry the same, we by this means do obtain a dry graduating Water, which being molten with any white or red metals makes them yield good Gold, and LUNE

on the Cupel to the great profit of the Artist.

43. Oil of Vitriol mixed with SAL ARMONIACK, is also of good use to bring over Tinctures, but not in that quantity as Butter of Antimony.

44. Our dry, sweet universal tinging water dissolves white Pebbles and Crystals, and changeth the same into precious Stones of several colours, excepting only their hardness, which it cannot communicate.

45. The easiest way to get the SOL or LUNE that is in JUPITER, is by casting it upon molten VENUS, which draws the SOL and LUNE to it self out of the JUPITER.

46. In like manner doth the REGULUS of Antimony when in flux readily draw to it the SOL and LUNA in JUPITER, and then washing the REGULUS with Niter we get the SOL and LUNA contained in JUPITER.

47. But this operation ought not to be done in Crucibles made of common Earth, which easily break and spill the metal, but in those that are made of a fat crucible Clay, mixed with coal dust, as is taught in the fifth part of my Furnaces.

48. As the Sulphur in Tartar coagulates a thin water into a hard Hepar, or Liver so called, so likewise doth a fixt Sulphur coagulate MERCURY into SOL and LUNA.

49. The often calcining of Salts and dissolving them in Water, doth purifie them, and makes them easily fluxible, and in particular Vitriol may by this means be so purified, as to yield its Oil with a very small heat.

50. The Solution of SATURN and LUNE poured into a volatile Spirit of MARS or VENUS, draws the tinging volatile Sulphur out of the Water to it self, and makes

the same Corporeal and fixed.

51.   Tartar contains a coagulating and tinging Sulphur, for it coagulates Water into a Hepar, and tinges red metals to a white Stone, which may be pulverized; this Sulphur is the cause why Tartar will not dissolve in cold Water as other Salts.

52.   When therefore the Tartar is freed of this Sulphur that coagulates all Water, then much good may be done therewith as well in Physick as Alchemy, and many other Arts besides.

53.   Tartar by being boiled in a strong Lye, lets go its coagulating Sulphur, and a neutral Salt proceeds from them both; but if we have a mind to separate the Lye from the purified Tartar we must do it with an acid, that may mortifie the Lye, by which means the purified Tartar will be left snow white.

54.   If the Lye be killed with a Spirit of Niter or AQUA FORTIS, then from both these contraries proceeds a good Niter.

55.   But if we make use of a Spirit of Salt, then there is made up of both a tartarized Spirit of Salt.

56.   If we take distilled Vinegar to mortifie the Lye, then from the joining of those both proceeds a neutral volatile Salt which is a good Diuretick in the Gout and Stone.

57.   This is the best way of purifying Tartar, which after this Operation is of far greater use in Physick and Alchemy than the common Tartar.

58.   Tartar as hath been said, contains a coagulating and tinging Sulphur, coagulating all Water

into a thick Hepar, and exalting the Colours of metals. Thus we see that by boiling Golden or Silver Vessels with Tartar, their several Colours are exalted.

59.   And whosoever has the Art of separating this Sulphur from Tartar, may by means thereof effect great and wonderfull things.

60.   A like wonder working Sulphur is likewise found in Animals, and more especially in man, who brings it with him into the World.

61.   Whence some Philosophers tell us, that ADAM brought the Philosophers Stone with him out of Paradise, and after his death carried it with him into his grave.

62.   Minerals also afford the same coagulating fixing and tinging Sulphur, for which reason the Philosophers Stone is said to be Animal, Vegetable and Mineral, because of each of these three Kingdoms an Universal medicine may be made for men and metals.

63.   But the easiest way is, when we extract the best part of all these three Kingdoms, and conjoin them according to Art for an Universal medicine.

64.   Wine is the chief of Vegetables, Man of Animals, and Gold of Metals.

65.   Spirit of Wine purges and purifies all things, with its purifying Flame; as may be seen in my Purgatory of the Philosophers.

66.   The volatile Salt of Animals, and especially of Man purifies all things by its volatilizing Virtue, as appears in our most secret SAL ARMONIACK.

67.   The Incombustible Sulphur of metals tingeth the Bodies of men and metals, to the highest pitch of

Health, as may be seen in the third APPENDIX to the seventh part of my PHARMACOPAEA SPAGYRICA.

68.   DEMOGORGON with his Russet mantle and green Coat, is the Grandfather of all the Heathen Gods i. e. of all metals.

69.   And like as in the Earth he doth generate and bring to perfection all metals, so also out of the same, if the Artist knows how to manage him, he perfects all unripe and imperfect metals, in a short time, with the help of Fire, to that degree that they shall endure the Test as well as SOL and LUNA.

70.   This wonderfull virtue of fixing all volatile minerals, the Philosophers call their secret Fire or proper Agent, wherewith not only the imperfect metals, as SATURN, JUPITER, VENUS, and MARS, but also volatile MERCURY, combustible common Sulphur, Antimony, Orpiment, and Arseniok may be fixed,, so as on the Cupel to leave SOL and LUNA.

71.   And as this Demogorgon, or invisible secret Fire of the wise Men, doth fix the unripe minerals and metals into LUNE and SOL; So likewise can it fix the said minerals and metals, and exalt them to an higher degree than that of SOL, even to the PLUSQUAM perfection of true Tinctures, whereby all imperfect metals may be changed into
SOL.

72.   This our Demogorgon hath the virtue even as it comes raw and unprepared out of the Earth to change and meliorate all metals as follows.

73.   It makes SATURN as hard and white as LUNE,

when tinged with it, of which all manner of Vessels and Dishes may be made, it only wants the sound of LUNE and enduring of the Test.

74. If a little of this Tincture be cast upon fluxed VENUS it presently becomes white and hard as Steel, continues as fusible as before, and yet is so hard that it cannot be filed, so that several vessels may be made of it, not subject to bending or breaking.

75. When cast upon melted JUPITER, it makes it hard as LUNE, and sounding like it, is of great use to make all sorts of Vessels of.

76. And amongst other things that may be made of it with great profit, this is one, viz, that Looking-Glasses may be made thereof, which being polished continue a long time clear and fair, without being obscured in moist Weather, as other metalline Glasses are, and all this by reason of the extraordinary hardness of the metal. SEE MY TREATISE OF LOOKING-GLASSES.

77. This Tincture cast upon LUNE, makes the same Coal-black throughout, so that it is no more like LUNE, of which Bells and Clocks may be made of a far better and clearer sound than those that are made of VENUS and JUPITER.

78. By this means also in times of War, or other danger LUNE may so be disguised as not to be known for such, and so may be a good way to preserve it from being taken by the Enemy.

79. In like manner it makes SOL so hard that it can no way be bent or destroyed, and therefore might be

of good use in many of the following cases.

80.   It would be very proper for some great Emperour or King to make his Statue of, it being indestructible, and not to be diminished or injured by any way whatsoever.

81.   Money coined of this SOL would be of good use if a King or Prince had a mind that his Coin should not be transported elsewhere, because differing so much from common SOL it would not be passible in other Countries.

82.   This golden Coin also would not be subject to be clipt or filed.

83.   Medals also might be made of this SOL, and would be a great curiosity besides the indesectibleness of them.

84.   It would be excellent also to make Rings of, especially such as are designed for the remembrance of Friends, as lasting for ever.

85.   It would be very proper to cast Seals of, or the divine Character or other secret SIGILS. SEE MY TREATISE OF THE DIVINE CHARACTER AND SEAL OF GOD.

86.   Or the said Divine Character being exprest upon my LAPUS IGNIS, which being but for a little while carried in ones mouth, cures many grievous Diseases without any other Medicine, might be set in this hard SOL, and so without wasting be carried constantly about one. SEE MY TREATISE OF THE MINERAL SQUILLA IN ORDER TO A LONG LIFE.

87.   Great Princes also might have Armour and Arms made of this hardened SOL. which would be much better than any of Iron or Steel, which easily take rust, to

which SOL is not Subject.

88. Of this SOL might also very conveniently be made Candlesticks and Lamps, with other Vessels for the use of the Church and Altar.

89. To many more uses this SOL might be put, especially for that by reason of its hardness, if suffers it self to be polished to that degree, as to cast a great lustre from it, like the Sun.

90. As to the further use hereof, SEE MY TREATISE de tribus Lapidibus ignium secretorm.

91. With the hardened LUNE, VENUS, JUPITER, MARS and SATURN, many profitable and curious things may be done, which for brevity sake are here omitted.

92. The Sulphur of the Philosophers when set free from his dark Prison, wherein he is detained by his Breathren, by our Key that opens all Locks, gives his Deliverer for reward, the possession of the three Kingdoms in the World, viz, enabling him to make all Vegetables grow swiftly; and very fruitfull, to cure the Diseases of all Animals, and to meliorate and exalt all Metals.

93. And when the Philosophers; saith SENDIVOGIUS, see this Sulphur restored to liberty, swimming in their Sea, they worship it, and draw it out with a Silver Line' though others do it with their SOL attracting Magnet and fix it into an universal Medicine, wherewith they afterwards effect wonders; AS MAY BE SEEN IN MY ELIAS ARTISTA, AND PURGATORY OF PHILOSOPHERS.

94. The Philosophers say, except first you make our SOL (that is, the redeemed Sulphur) and Mercury

white you'll never be able to make them red.

95. They say also our SOL tingeth not except it be first tinged it self, that is exalted in its colour.

96. All things in the World have their rise from Fire and Water, and derive their Purity or Impurity from the Purity or Impurity of their Parents.

97. The common Fire brings forth its fruits very slowly, whether they be stones, Minerals, Animals, or Vegetables

98. And so do likewise the warm and dry Sun, and moist Earth; but when we assist Nature with Art; then she works much more swiftly and brings her Fruits to maturity in much shorter time.

99. The Meteors in the Firmament which are made of Fire and. Water, especially Thunder and Lightning produce someti.mes Stones, and cast them to the Earth.

100. A common fulminating Powder made of Sulphur, Niter and Tartar gives a stinking offensive smoke, corrupting some things, and meliorating others; whereas a Fulmen prepared of Niter, JUPITER and MERCURY, yields a particular tinging mercurial Water. The Fulmen of VENUS tinges MARS into Copper, that of LUNE graduates VENUS into LUNE and the Fulmen of SOL graduates and tinges MARS into SOL.

The universal Fulmen of the great Tincture graduates all Metals into SOL, which God of his mercy grant unto us, AMEN.

THE COMPLETE WORKS
OF

RUDOLPH
GLAUBER

trans: Chris. Packe

RAMS
1983

THE CENTURYS

FIFTH

THE FIFTH CENTURY

The best particular and chiepest Universal.

1. When with the help of SENDIVOGIUS his CHALYBS, or GLAUBER'S Magnet, we have extracted the colour from SOL, and again restored it through VENUS and Antimony, we may by oft repeating the said extraction and restoration get great profit, this being one of the best particulars that can be. This multiplication of SOL may very well be compared with the generation of Man for as a Man in generating, doth with meat and drink restore the loss of his Seed, by which means he continues the said multiplication for a long time, by turning the meat he eats into Prolifick Seed; so likewise the Chymist changeth VENUS, MARS, JUPITER, SATURN, MERCURY and LUNE into SOL, by feeding the dis—spirited SOL that has lost its colour with them, restoring it to its former strength and vigour.

2.	The Sperm of Man is not the Seed of Man, but only the Shell and receptacle thereof, as may be seen in Old Men, whose Sperm is unfit for generation by reason of the weakness of their vital Spirit.

3.	So likewise the Seeds of Vegetables, are not all pure Seed, but the house and Vehicle thereof, that is, of the growing and multiplying virtue, which appears in that when the Seeds have been kept so long till this vital virtue is exhaled from them, they never bring forth any thing.

4. No more can SOL be said to be the Seed of

metals; but only the receptacle thereof, for the Seed is not the whole Body, but only the lively colour of the Body, and the vegetative and multiplicative virtue that is hid in it.

5.   Now as the Seed of Vegetables is more perfect and noble than the Vegetables, so likewise is mature fixt SOL, more perfect than MERCURY, SATURN, JUPITER, VENUS, MARS, though in the imperfect metals also a Seed be hid, but not so fixt and good as that in Gold.

6.   The imperfect metals may be compared to an Herb; whose Seed is not yet ripe, which being put into the ground cannot grow or multiply, but rots in the Earth.

7.   The virtue of Corals lies not in their whole Bodies, but in their colour; and therefore PARACELSUS bids us not to make use of Corals in substance, but extract their Tincture, and use that for Physick, wherefore also he rejects white Corals, as being an unripe Fruit, from any use in Physick.

8.   For this reason also the immature grey Pearls, which are frequently found in Cockle Shells in fresh running waters, are looked upon as useless in Physick.

9.   And this not without reason, for as unripe Grapes are the cause of griping of the Guts, and hurt the Body; so ripe Grapes nourish and strengthen the same, especially when by fermentation they have quitted their Faeces.

10.   All imperfect metals subvert and trouble the Stomach, and cause vomiting and purging, and that by

reason of their unripeness.

11.　Whereas on the contrary SOL taken into the Body causeth not the least alteration, but powerfully strengthens the same when reduced to Potability.

12.　Thus SOL may be compared to ripe Grapes; which when eaten raw, do indeed no hurt to the Body, but rather affords some nourishment, yet cannot strengthen the Heart, Brain, and whole Body, and make a cheerfull mind; but when by fermentation they are delivered from their skins and other impurities; they readily and as it were ma moment perform this.

13.　In like manner when SOL by fermentation hath laid aside his gross Body and become Spiritual, if then made use of, it not only nourisheth as ripe Grapes; but exerts its virtue like a Spirit or QUINTESSENCE of Wine, penetrating the whole, and making it lively, strong and vigorous throughout.

14. Neither do the other metals display their hidden virtue, until by fermentation and distillation; they be subtilized and their gross Bodies laid aside.

15.　Thus when LUNE by fermentation and distillation is subtilized, then it draws away all Diseases of the Brain, and corroborates the same exceedingly even as SOL doth the heart.

16.　VENUS so purified strengthens the Reins and procreative faculty.

17.　The volatile sweet Spirit of MARS, removes all obstructions whatsoever, provokes the terms in Women, and opens the Haemorroides in Men.

18.　The sweet Spirit of SATURN cures all inward

and outward hot Distempers.

19.   The sweet Spirit of JUPITER cures all Distemper's of the Lungs.

20.   The volatile Spirit of MERCURY cures the venereal Distemper.

21.   N. B. These volatile spirits of metals must be cautiously used, as being of very great force.

22.   The manner of preparing them, may be seen in my Book of Fires, but most plainly set down in my description of the most secret SAL ARMONIACK.

23.   All Spirits act according to their nature and property either good or ill, as the Bodies are good or evil from whence they are taken.

24.   THE SPIRIT QUICKENS, THE BODY OR FLESH PROFITS NOTHING, SAITH CHRIST, John 6.

25.   These words are ill interpretated, when understood by some, as if Spirits only were of use,and Bodies not at all, which is a great mistake, as it is applied by some.

26.   Indeed in Metals, Vegetables and all Animals without the use of reason, who grow, move, and live, by the driving of their in-born Spirit, it does hold true, for when their Spirits are by Art separated from their Bodies, the said Bodies are thence-forward of no use, as being upon the separation of their Spirit, dead and without all virtue.

27.   But the case is different with Man, who being created in the Image of God, and endowed besides his Animal Spirit, with an immortal Soul, which latter only and immediately derives from God, and not from nature,

as the mortal Spirits of Animals do.

28.   Wherefore PYTHAGORAS was much mistaken, in believing that the immortal Souls of Men, when departed from their Bodies did immediately enter into those of Beasts.

29.   Which mistake of his seems to have been occasioned hence, because he knew how by Art to take away the Soul, i. e., Tincture from SOL, and transfer the same to an imperfect metal, thereby making the same in all things like to true natural SOL.

30.   Certain it is that this may be done by art, for the fixt Body of SOL may be destroyed, its Soul extracted, and by being joined to another metal make it good SOL.

31.   When this disanimation of SOL is duely performed, the Body is left wholly dead, and is in all things like a volatile unmalleable mineral, and cannot endure the test, but fumes away like Arsenick with a little Fire.

32.   But in case this disanimating of SOL be not rightly done, so that the Body continues as white asLUNE, and malleable (which is a sign that some life is still left in it) then his Colour may be restored again by means of imperfect minerals, as well as his former fixedness in the Fire.

33. But when the Body of SOL will no longer endure the fire, but goes away in smoke, then we can say it truly is dead and no more SOL.

34.   He that finds difficulty to believe this, let him read PARACELSUS, SENDIVOGIUS and other Philosophers.

35.   SENDIVOGIUS saith, OUR STEEL, that is, our Magnet, CAN DRAW FROM THE RAYES OF THE SUN, WHAT MANY HAVE SOUGHT FOR AND NOT FOUND; IF THIS OUR MAGNET COPULATE ELEVEN TIMES WITH SOL, THE SOL BECOMES WEAKENED ALMOST TO DEATH, AND THE STEEL OR MAGNET SHALL CONCEIVE AND BRING FORTH A SON MORE ILLUSTRIOUS THAN HIS FATHER.

36.   From which words it appears that SENDIVOGIUS had the Art to disanimate SOL, else could never have writ so plainly concerning it.

37.   It is certain also that there are some, that at this time can do as much within a few hours, I having lately been an Eye-witness of the same, with three other persons in company.

38. It is not necessary to say any more how this cheap and speedy way of disanimating SOL is to be performed; forasmuch as all the Philosophers writings are full of it.

39.   However to pleasure the unskillfull I will add thus much; that this may be done four several ways.

40.   But the easiest and chiepest way is by means of Spirit of Wine, and a microcosmical saline Spirit; yea this extraction may be performed by a Spirit of Wine alone, without any animal Spirit, or by an animal saline Spirit without the Spirit of Wine.

41.   If this were not so; we might have reason to accuse both ancient and modern Philosophers of falshood, who tells us that ADAM brought the Philosophers Stone with him out of Paradise, and after his death took it with him into his Grave.

42.   Which words may seem strange to some,

forasmuch as he was driven bare and naked out of Paradise.

43.   Yet the Authority of those who assert this being so great and incontestable it cannot well be called in question.

44.   What therefore the Philosophers meant by this Stone which ADAM brought with him out of Paradise; is well worthy our Enquiry.

45.   The Philosophers commonly say our stone is a stone and no stone, & etc. which implies thus much, that to outward view it is a stone, but in deed and in virtue a Concentrate form of SOL.

46.   Wherefore PETRUS BONUS saith, WE DO NOT SEEK SOL, BUT THE FORM OF SOL.

47.   What then is properly this form of SOL?

48.   Answer. It is a substance which to outward view looks like a contemptible stone and yet is of such superlative Virtue, that when joined with imperfect metals on the Fire it transmutes them into the highest perfection of SOL.

49.   It may further be demanded, whether ADAM brought such a matter with him out of Paradise, whereby this kransmutation of metals into SOL may be performed?

50.   Answer. Yes he did bring such a matter with him out of Paradise, and after death took it with him to his Grave wherewith all Diseases of mankind may be cured; and all metals changed into the finest SOL.

51.   If this be so, might some say, ADAM must either have been very blind; in not discerning the Treasure he was possessed of; or very envious in not

communicating the same to his Posterity.

52.   I cannot believe that ADAM, out of envy withheld this secret from his Children, but rather suppose that the blindness into which his fall had cast him, was the cause of his not perceMng the great Jewel he had about him.

53.   But how could he be blind, who was made by God himself; and after his own Image?

54.   ADAM was certainly blind, and his blindness proceeded from his Pride because he aspired to be like to God; he was not blind as to his outward Eyes; but his heart was blinded, which is by far the worst blindness of the two, For all sin and wickedness blinds the hearts of men, that they cannot perceive the folly of their doings.

55.   Thus ADAM by also means of his disobedience of God; became so blind, as not to perceive, or be sensible of the Love that God had for him before his Fall, and how righly he had endowed him.

56.   Whence also his Children were so wicked and blind, that the one Brother slew the other; which wickedness hath still encreased in their Posterity; as appears by the deluge and the destruction of SODOM AND GOMORRAH.

57.   And thus the World from day to day still grows worse and worse; notwithstanding the Examples of God's Vengenance against Sinners.

58.   And all this proceeds because men are so generally blinded by the Devil in sin and wickedness.

59.   But to leave this it may be further

questioned; that seeing all mankind is become so blinded through ADAM'S fall, as not to discern the Jewel they carry about them; who then was the Person that first discovered; that man was the Possessour of so great a Treasure?

60.   Answer. Who was the first discoverer of this Treasure I cannot tell, but thus much is certain; that it was an honest man and fearing God; because God doth not reveal his secrets to the wicked; wherefore THO. AQUINAS saith, OUR HOLY ART, EITHER FINDS A MAN HOLY, OR MAKES HIM SO.

61.   But some will say don't we read of Heathens that have been Possessours of the Philosophers stone; and how can we imagine that those who have no knowledge of God; and are blinded with sin, should ever be able to find out so great a Mystery?

62.   Answer. Those Heathens that have been Possessors of this great secret; were not without the knowledge of God; for they lived according to the Law of Nature, honouring God and loving their Neighbour; wherefore also God accepted of them. They learnt to know God from his Works of Wonder, and according to their knowledge, loved, honoured and feared him; and so were made Partakers of his grace, light; and the knowledge of his secrets.

63.   We are also to know that the Ancient Philosophers know more than one way to attain the Philosophers Stone; though indeed the most of them sought it in minerals and metals; which is the longest way.

64. And that because it is impossible to change the metallick species; without bringing metals back into their first matter. SEE MY TREATISE OF THE PRINCIPLES OF METALS, AND THE SEVENTH PART OF MY PHARMACOPOEA SPAGYRICA.

65. But others have taken a nearer way to attain this secret; and to some Christians God hath been pleased to discover the shortest way of all; by revealing unto them that he made ADAM every way perfect; gMng him all that was necessary, either for his Soul or Body.

66. Now that ADAM could not discern how richly God had endowed him, was his own fault, because he was disobedient to God, following the deceitfull Serpents advice:

67. And after this manner doth the Devil yet daily deceive Men, by perswading them to do against the commands of God; and that their disobedience shall not bring any mischief upon them as God's threatnings seem to import.

68. This then is the reason why Men do not understand the secrets of God, because they give too much way to sin, whereby they become blinded, that they can neither see not hear the good that comes from above.

69. Now the reason why most of the Alchemists have sought for this great gift of God in Minerals and Metals, and especially in SOL, is this, because their intention was to multiply SOL, which they supposed could no way better be done than by sowing it like other Seeds in the Earth, but could not imagine that besides common

SOL, there were other subjects, wherein the SOL-making virtue did reside.

70.   Which opinion of theirs was probably grounded upon that saying of the Philosophers. WHAT YOU SOW, THAT YOU SHALL REAP.

71.   This seems at first sight very rational, that from filth or excrements no good, and so no SOL can come; but let us hear the other side also, and we shall be otherwise informed.

72.   For the Philosophers say that their Medicine is Vegetable, Animal and Mineral; so that Vegetables and Animals are not excepted.

73. ALBERTUS MAGNUS, writes that the greatest mineral aurifying virtue is in Man, and especially in his Head between his Teeth, and proves it; because in dead Men's Skulls he had found grains of SOL sticking between the Teeth.

74.   The same is also confirmed by THOMAS AQUINAS, RHASIS, JANUS LACINIUS, and others.

75.   There is also an old Book, whose Authour is unknown, which treats at large of that subject which ADAM brought with him out of PARADISE, wherein the Operator is warned to have a great care of the fumes of the matter as he would avoid the Plague, or the most deadly Poison, From this Authour I have alledged some passages in my other Writings, and shewed that the Philosophers Stone may be prepared of any subject whose Elements may be separated.

76.   Now certain it is that from all Animals and Vegetables, the Elements may be separated, and

consequently follows, that from all Vegetable and Animal Subjects, the Philosophers Stone, or universal Medicine for the Bodies of Men, and Metals may be prepared.

77.   MORENUS ROMANUS, who prepared this Medicine for King CALID, declares that he took the subject matter of it from Man.

78.   For when the King asked MORIENUS, in what kind of subject the Philosophers Stone was to be lookt for; he answered, the Medicine is in thy self O King; wherefore also after that he had finished the Work, he wrote round about the Glass, in which the Medicine was, these words: HE WHO CARRIES ALL ABOUT HIM, NEEDS NOT THE HELP OF ANOTHER.

79.   Thereby intimating, that he always carried about with him, whatsoever was necessary for the preparing of the Medicine, and therefore did not stand in need of the King's assistance.

80.   This same honest MORIENUS, writes plainly concerning the preparation of this Medicine, and doth as it were with his finger point us to the matter, in these words of his, quoted by ARNOLDUS DE VILLA NOVA; GRIND THE PHLEGMATICK AND CHOLERICK WITH THE SANGUIN, UNTIL IT BECOME A TINGING HEAVEN, & etc.

81.   ARNOLDUS explains these words of MORIENUS thus: THE PHLEGMATICK IS COLD, AS MERCURY, THE SANGUIN IS WARM AND MOIST, AS THE SOL OR GOLD, THE CHOLERICK IS HOT AND DRY, AS SAL ARMONIACK: intimating that of these three, MERCURY, SOL and SAL ARMONIACK, the Philosophers Stone is to be prepared.

82.   But that he meant not this concerning common

MERCURY, SOL, and SAL ARMONIACK is apparent from this, that MORIENUS, as soon as he had prepared the Medicine for the King, went away prMtely, without expecting any reward from the King; it also appears from the answer before mentioned, which he made to the King, that he spoke of such a MERCURY, SOL and SAL ARMONIACK which every Man carries about with him.

83. This is abundantly confirmed by all the Philosophers that went his way, forasmuch as they declare that no charges are required to the preparation of it, that their subject is a contemptible matter cast out upon Dunghills, and trod under feet, and that the Poor have it as well as the Rich.

84. MORIENUS yet more clearly intimates this, in telling us that the matter whilst it is preparing, exhales a smell like to that which comes from the Graves of the Dead, which is a very offensive smell.

85. Now like as Vegetables whilst they are putrifying give forth an ill scent; and Animals a worse, as appears in the stink of rotton Eggs, and the putrefaction of Man's Blood, especially when the same are putrified in a close Glass in warm Horse Dung.

86. For without putrefaction, there can be no separation of the Elements by Distillation, and if no separation be made, neither can any melioration or exaltation be expected.

87. We know that every CHAOS, as it is a product of the four Elements; contains many impurities, and in particular much dead Earth, and Water void of all virtue; and that the Element of Fire alone is proper to

heal and meliorate Men and Metals.

88.   Wherefore seeing that no separation of the Elements can be without a foregoing putrefaction we must conclude putrefaction to be the beginning of our Work, without which no good can be expected.

89.   Now he that knows our Horse Dung, and how to putrifie the well known and every where to be found most universal natural subject by means of the same, he will easily afterwards by Distillation separate the most pure and all things penetrating and meliorating Element of Fire, from the gross Chaos to his great satisfaction, and make use of the same to the astonishment and wonder of the ignorant.

90.   But in this state it is only good for the health of Man; and therefore in order to its meliorating of Metals, the pure Element of Fire must be first fixed with SOL, by which means it obtains ingress into imperfect Metals, reducing them to the perfection of SOL.

91.   Now, when the pure Element of Fire is separated from the Chaos, and reduced to the highest degree of purity, then it stinks no more, neither is poisonous as it was before purification, but is an Antidote against all poisons whatsoever, wherefore also the Philosophers have called their Medicine THERIACA.

92.   But all this is to be understood only of that subject which every man carries about with him; and brings with him out of his Mothers Womb.

93. If any one following the Letter of MORIENUS, should take for his subject common MERCURY, SOL and SAL

ARMONIACK, neither will he be mistaken, but if he rightly proceeds will have a good Work, though it be not at all necessary to make use of common SOL and MERCURY, because our natural subject contains both a living SOL and MERCURY.

94. It is no prejudice to our Animal Subject, if we join Minerals with it, because our SOL joins it self with all subjects, and unites it self readily with them. But if we be ignorant of the due proportion and composition of SOL, MERCURY, Sulphur, or any other metal or mineral, then it is better to prepare our Medicine out of this one subject only, because so there is less danger of erring, as I can witness by experience.

95. I have also found by experience that this Microcosmical Subject is alone sufficient, without the addition of any minerals or metals, to meliorate all imperfect metals.

96. As to particular this of all others hath pleased me best, viz. RECIPE common SOL, and with the help of our Magnet disanimate it so, that it may be no longer SOL, as not enduring the Test, and smokeing away with a small Fire like Arsenick.

97. Then take this SOL and conjoin it with our Microcosmical Subject, with which digest a solution of LUNE, by which means the LUNE will be meliorated, and on the Cupel leave SOL to good profit.

98. But if we join common MERCURY and common SOL with it, and cast this mixture into a solution of MARS, and digest it for some days then the pure SOL and easily flowing MERCURY graduates a good part of the gross and

difficultly flowing MARS into good SOL, to the great satisfaction of the Artist.

99.   And if we unite LUNE and JUPITER therewith, and cast this mixture into a solution of VENUS, and digest it the moist way, then by means of our secret Salt these two white united metals change the red VENUS with little loss of weight into good LUNE that will abide the Test. And it is indeed matter of wonder, that our universal Salt, should be of so great virtue, when fermented with white or red metals, to change other imperfect metals into good SOL and LUNE on the Test.

100. Wherefore this shall be my conclusion, that in Man is hid the greatest virtue of changing all metals, as well as the Bodies of Men, both universally, and particularly; which if intended for the melioration of metals, the adding of fixt SOL and LUNA for a ferment will facilitate the ingress into other metals, and further diffuse its tinging virtue.

N.B. I shall not be satisfied till I have given a fuller and plainer description of this Royal Labour, which I intend to do in the sixth Century, If God permit.

NOTE. In the collected writings of Glauber, there is no sixth Century; in additions of other writings he mentions; I suppose they are somewhere still in Latin or German. D.H.

A Word from the Publisher

Thank you for purchasing this small work from The R.A.M.S. Library of Alchemy. During his lifetime, Hans Nintzel was dedicated to the identification, acquisition, study, retyping and, when necessary, translation of what he considered to be the most important known works on Alchemy. Hans was assisted by his sparse network of fellow Alchemists, all members of the Restorers of Alchemical Manuscripts Society (R.A.M.S.). I was an active member of R.A.M.S.

My goal is to publish all of the works originally made available through R.A.M.S. as photocopies. To facilitate this, I have chosen to have the books professionally printed. I also have a few titles that I intend to add to the original R.A.M.S. Library, selected by strict criteria established by Hans.

If you have a work on Alchemy that you believe should be a part of the R.A.M.S. Library, please contact me through R.A.M.S. Publishing Company.

Philip N. Wheeler